IIW Collection

Series editor

IIW International Institute of Welding,
ZI Paris Nord II, Villepinte, France

About the Series

The IIW Collection of Books is authored by experts from the 59 countries participating in the work of the 23 Technical Working Units of the International Institute of Welding, recognized as the largest worldwide network for welding and allied joining technologies.

The IIW's Mission is to operate as the global body for the science and application of joining technology, providing a forum for networking and knowledge exchange among scientists, researchers and industry.

Published books, Best Practices, Recommendations or Guidelines are the outcome of collaborative work and technical discussions - they are truly validated by the IIW groups of experts in joining, cutting and surface treatment of metallic and non-metallic materials by such processes as welding, brazing, soldering, thermal cutting, thermal spraying, adhesive bonding and microjoining. IIW work also embraces allied fields including quality assurance, non-destructive testing, standardization, inspection, health and safety, education, training, qualification, design and fabrication.

More information about this series at http://www.springer.com/series/13906

Erkki Niemi · Wolfgang Fricke
Stephen J. Maddox

Structural Hot-Spot Stress Approach to Fatigue Analysis of Welded Components

Designer's Guide

Second Edition

Erkki Niemi
Department of Mechanical Engineering
Lappeenranta University of Technology
Lappeenranta
Finland

Stephen J. Maddox
The Welding Institute
Cambridge
UK

Wolfgang Fricke
Institute of Ship Structural Design and
Analysis
Hamburg University of Technology
Hamburg
Germany

ISSN 2365-435X ISSN 2365-4368 (electronic)
IIW Collection
ISBN 978-981-13-5429-8 ISBN 978-981-10-5568-3 (eBook)
DOI 10.1007/978-981-10-5568-3

Printed on acid-free paper

This Springer imprint is published by Springer Nature
The registered company is Springer Nature Singapore Pte Ltd.
The registered company address is: 152 Beach Road, #21-01/04 Gateway East, Singapore 189721, Singapore

Preface

This 'Designer's Guide' follows previous recommendations published in 1995 by the International Institute of Welding (IIW), 'Stress Determination for Fatigue Analysis of Welded Components' edited by E. Niemi. It represents the latest product of the continuing joint-activity of Commissions XIII (Fatigue) and XV (Design) to develop better fatigue design and analysis procedures for welded structures. The Guide focuses on one particular aspect, namely the hot-spot stress approach. It is intended to provide practical guidance on the application of that approach, based on the current state-of-the-art. However, it is also hoped that it will promote wider use of the approach, assist code-writers in its introduction into design Standards and encourage further research.

A shortcoming of current fatigue design rules for welded components and structures is that they have not kept pace with computing developments in design, notably stress determination by finite element analysis (FEA). The basic design method embodied in the rules was actually developed over 30 years ago, when computers were something of a novelty and structural analysis relied mainly on the use of standard formulae and experience. Thus, it was entirely reasonable to base fatigue design on nominal stresses, as is currently the case. However, computer-based analyses like FEA are now used routinely and, with increased computing power, their capabilities are increasing. Thus, general guidance is needed on the determination of the stresses to be used in conjunction with the current nominal stress-based design rules. The IIW has been particularly active in developing such guidance, as noted above. However, this represents an interim solution in that there is also scope for new design methods that take more advantage of the potential output from FEA. One such approach, for designing weld details from the view-point of potential failure from the weld toe, is the so-called *hot-spot stress method*. This makes use of the stress adjacent to the weld that includes the stress concentration effect of the welded joint, but excludes the local notch effect of the weld itself. It is generally referred to as the structural, or geometric, stress and its value at the weld toe is termed the *structural hot-spot stress*. It is then used in conjunction with appropriate S-N curves representing the notch effect of the weld toe.

The hot-spot stress approach should lead to increased accuracy in fatigue design. The current nominal stress-based design curves include some allowance for the stress concentration effect of the weld detail, but in a rather crude way. That stress concentration effect is known to depend on the dimensions and proportions of the weld detail, but the fatigue data used to establish the design curves were obtained from a variety of test specimens with varying geometry. FEA enables the actual geometry and dimensions to be modelled, leading to a more precise estimate of the stress concentration due to the weld detail.

Although in use for tubular structures for over 25 years, only tentative guidance is available on application of the structural hot-spot stress method to plate structures. Key requirements are:

(a) The definition of the structural hot-spot stress and how it is obtained from stress analysis;
(b) The choice of hot-spot stress design *S-N* curves.

Both issues are addressed in the present document, but especially the first, on the basis of the current state-of-the-art. It reflects the findings of major joint-industry research programmes, as well as deliberations by IIW and other standardizing bodies. However, research continues and it is anticipated that improvements to the procedures described will be made in future.

The document is the product of several years of study by IIW Commission XIII Working Group 3, 'Hot-spot stress method in fatigue analysis of welded components', initially under the Chairmanship of Prof. Erkki Niemi (Finland), but then under Prof. Wolfgang Fricke (Germany) when Prof. Niemi retired. Final drafting of the document was undertaken by Prof. E. Niemi, Prof. W. Fricke and Dr. S.J. Maddox (UK). Members of both Commission XIII and Commission XV have participated in the work, as well as practicing engineers from industry. However, special acknowledgement is made to the following:

Prof. G. Marquis—Finland
Dr. M. Huther, Dr. H-P. Lieurade, D. Turlier—France
Prof. A. Hobbacher, Prof. H. Petershagen, Prof. C.M. Sonsino—Germany
Prof. C. Miki, Mr. I. Kojima—Japan
Prof. P. Haagensen, Dr. I. Lotsberg, Dr. K.A. MacDonald—Norway
Prof. J. Samuelsson—Sweden
Dr. P. Dong—USA.

In 2013 it was decided to revise the document in order to consider recent developments in the field. In particular, more emphasis was placed on the structural stress determination by through-thickness stress linearization, on the alternative approaches by Dong, Yiao/Yamada and Haibach and on additional examples to demonstrate the application of the recommendations. The revised version was finalized in early 2016, again with discussion and comments from the aforementioned Working Group and Commissions of the International Institute of Welding, which is highly appreciated.

Lappeenranta, Finland — Erkki Niemi
Hamburg, Germany — Wolfgang Fricke
Cambridge, UK — Stephen J. Maddox

Contents

Abstract

This document provides background and guidance on the use of the structural hot-spot stress approach to the fatigue design of welded components and structures. It complements the IIW recommendations for "Fatigue Design of Welded Joints and Components" and extends the information provided in the IIW recommendations on "Stress Determination for Fatigue Analysis of Welded Components". This approach focuses on cases of potential fatigue cracking from the weld toe. The approach has been in use for many years in the context of tubular joints. The present document concentrates on its extension to structures fabricated from plates and non-tubular sections.

Following an explanation of the structural hot-spot stress, its definition and its relevance to fatigue, methods for its determination are described. Stress determination from both finite element analysis and strain gauge measurements is considered. Also, parametric formulae for calculating stress increases due to misalignment and structural discontinuities are presented. Special attention is paid to the use of finite element stress analysis and guidance is given on the choice of element type and size for use with either solid or shell elements. Design *S-N* curves for use with the structural hot-spot stress are presented for a range of weld details. Finally, use of the recommendations is illustrated in four case studies involving the fatigue assessment of welded structures using the structural hot-spot stress approach.

Chapter 1
Introduction

1.1 General

Traditional fatigue analysis of welded components is based on the use of nominal stresses and catalogues of classified details. A particular type of detail is assigned to a particular fatigue class with a given *S-N* curve. Such a method is used in the IIW fatigue design recommendations [1]. This nominal stress approach ignores the actual dimensional variations of a particular structural detail, which is an obvious drawback. Moreover, the form of a welded component is often so complex that the determination of the nominal stress is difficult or impossible. This is true even if the finite element analysis (FEA) method is used for the stress analysis.

In the context of potential fatigue failure by crack growth from the weld toe or end, the structural hot-spot stress approach goes one step forward. Here the calculated stress does take into consideration the dimensions of the detail. The resulting structural stress at the anticipated crack initiation site ('hot-spot') is called the structural hot-spot stress. Structural stress includes the stress concentrating effects of the detail itself but not the local non-linear stress peak caused by the notch at the weld toe. This notch effect is included in the hot spot *S-N* curve determined experimentally. This is reasonable because the exact geometry of the weld will not be known at the design stage. The variation in the local geometry of the weld toe is one of the main reasons for scatter in fatigue test results. By using the lower-bound characteristic *S-N* curve, lower bound quality of the weld toe is incorporated into the analysis. A single *S-N* curve should suffice for most forms of structural discontinuity, providing the weld toe geometry is always the same.

An obvious reason for introducing the structural hot-spot approach is the availability of powerful computers and software, which make detailed FEA possible for most design offices. However, the approach is also a valuable tool for choosing

E. Niemi et al., *Structural Hot-Spot Stress Approach to Fatigue Analysis of Welded Components*, IIW Collection, DOI 10.1007/978-981-10-5568-3_1

the locations of strain gauges when validating design by field-testing prototype structures. Moreover, finite element analyses make it possible to produce parametric formulae in advance for easy estimation of structural stresses at various hot-spots.

The hot-spot approach was first developed for fatigue analysis of welded tubular joints in offshore structures. Corresponding fatigue design rules were published by the American Petroleum Institute, the American Welding Society, Bureau Veritas, UK Department of Energy, etc. A review of this topic can be found in Ref. [2]. There is now an increasing demand for application of the approach to be extended to all kinds of plated structures. Some progress has been made in doing this, but at present there are differences in the methods recommended for estimating the structural hot-spot stress.

The first general design rule to include the structural stress (referred to at the time as the geometric stress) approach was the European pre-standard ENV 1993-1-1 [3] (Eurocode 3) but only limited guidance was given. Later, the International Institute of Welding (IIW) published new recommendations containing four fatigue design approaches, including the hot-spot approach [1]. A background document was also published focusing on definitions and the determination of stresses used in the fatigue analysis of welded components [4].

Subsequently, further research has led to improved procedures for determining the structural hot-spot stress, particularly using FEA [5, 6], and the provision of background fatigue test data from which to derive suitable design *S-N* curves [7–10]. Furthermore, the ability to establish through-thickness stress distributions using FEA has enabled a method to be developed that uses this information to calculate the structural hot-spot stress. Previously attention has focussed on use of the surface stress distribution, approaching the weld in question, and determination of the structural hot-spot stress using an extrapolation procedure. Use of the through-thickness stress distribution instead should avoid the need for extrapolation.

The goal of the present document is to help design engineers and stress analysts to implement the structural hot-spot approach in practice. Practical examples of the application of the methods described are given in the form of Case Studies in Chaps. 7–10. Moreover, the document should serve as a reference when detailed guidelines for design are developed for particular welded products. The recommendations given here are intended for design of general welded structures subjected to fatigue loading. The document is mainly focused on plated structures, such as bridges, cranes, earth moving machinery, ship hulls, etc. Specific rules are already available for certain fields of application, including tubular structures [11], ship hulls [12, 13], pressure vessels [14] and steel products in general [15].

In view of the scope of current experience and the availability of relevant fatigue test data, the recommendations presented in this document are only intended for plate thicknesses above 3 mm. However, the structural hot-spot stress has successfully been applied also to thinner plated structures, e.g. in the automotive industry.

1.2 Safety Aspects

When using this document, the general guidelines on fatigue analysis and safety aspects given in the latest version of the IIW Recommendations [1] should be taken into consideration. For example, it may be appropriate to multiply the fatigue loads by a partial safety factor, γ_F, which will normally be specified in the appropriate design code for a particular structure. Similarly, the fatigue strength may need to be divided by γ_M, as required.

It should be noted that the *S-N* curves presented in this document do not include any allowance for inaccuracies in stress determination. Furthermore, the stress raising effect of misalignment in welded joints and other such imperfections may not have been taken into account. If the stress analyst chooses to simplify the stress determination, for example by omitting the extrapolation of the structural stress to the weld toe or by totally neglecting misalignment, the load factor, $\gamma_{F,}$ should be increased accordingly.

References

1. Hobbacher, A.F.: Recommendations for fatigue design of welded joints and components, 2nd edn. Springer (2016)
2. Marshall, P.W.: Design of welded tubular connections. Basis and use of AWS code provisions. Dev. Civ Eng. **37**,412. Elsevier (1992)
3. ENV 1993-1-1. Eurocode 3: Design of steel structures–Part 1-1: General rules and rules for buildings. European Committee for Standardization, Brussels (1992)
4. Niemi, E.: Stress determination for fatigue analysis of welded components. Abington Publishing, Abington Cambridge (1995)
5. Fricke, W.: Recommended hot spot analysis procedure for structural details of ships and FPSOs based on round-robin fe analyses. Int. J. of Offshore and Polar Engng. **12**(1), 40-47 (2002)
6. Dong, P.: A structural stress definition and numerical implementation for fatigue analysis of welded joints. Intl. J. Fatigue **23**, 865–876 (2001)
7. Partanen, T., Niemi, E.: Hot spot *S-N* curves based on fatigue tests of small MIG-welded aluminium specimens. IIS/IIW-1343-96 (ex. doc. XIII-1636-96/XV-921-96), Welding in the World. **43**(1), 16-23 (1999)
8. Partanen, T., Niemi, E.: Collection of hot-spot *S-N* curves based on tests of small arc-welded steel specimens, IIW Doc XIII-1602-99, (1999)
9. Maddox, S.J.: Hot-spot stress design curves for fatigue assessment of welded structures. Intl. J. Offshore Polar Eng. **12**(2), 134–141 (2002)
10. Maddox, S.J.: Hot-spot fatigue data for welded steel and aluminium as a basis for design, IIW Document No. XIII-1900a-01, (2001)
11. Zhao, X.-L., Packer, J.A.: Fatigue design procedure for welded hollow section joints, Recommendations of IIW Sub-commission XV-E. Woodhead Publishing, Cambridge (2000)
12. G.L.: Rules for classification and construction, Part I, Ship technology, 1.1 Seagoing ships—Hull, Part V, Analysis techniques, edn 1998. Germanischer Lloyd, Hamburg (1998)

13. B.V.: Fatigue strength of welded ship structures. Publication NI 393 DSM R01 E. Bureau Veritas, Paris (1998)
14. Standard, European: prEN 13445-3, Unfired pressure vessels—Part 3: Design. European Committee for Standardization, Brussels (2009)
15. BS 7608:2014, Guide to fatigue design and assessment of steel products, British Standards Institution, London (2014)

Chapter 2
The Structural Hot-Spot Stress Approach to Fatigue Analysis

2.1 Field of Application

The structural hot-spot stress approach applies to welded joints for which:

- the fluctuating principal stress acts predominantly transverse to the weld toe (or the ends of a discontinuous longitudinal weld);
- the potential fatigue crack will initiate at the weld toe or end.

In Fig. 2.1, the hot spot approach applies to cases (a-e). For reasons that will be discussed later, it is also necessary to distinguish between joints that are fully load-carrying and non-load carrying. Cases (c) and (h) represent load carrying fillet welds, whereas case (b) represents non-load carrying fillet welds. Fillet welds at cover and collar plates (lugs), case (d) and (i), are actually partial-load carrying welds. Also end welds in cases (e) and (j), representing gussets or brackets welded on plates, are partial-load carrying. If there is any doubt about the choice of category, the joint should be assumed to be load-carrying. The structural hot-spot stress approach presented here is not applicable to cases where the crack will grow from the weld root and propagate through the weld throat, cases (f–j) in Fig. 2.1. Good design practice aims to avoid this kind of behaviour because the crack is not visible before it has propagated through the weld. However, approaches also exist for their assessment [1], which are partly classified as structural approaches. Moreover, the structural hot-spot approach does not apply to continuous welds subject to longitudinal loading. The nominal stress approach [2] is sufficient for such cases.

The weld detail being assessed will often be situated in a bi-axial stress field. Then it is usually sufficient to apply the structural stress approach to that principal stress which acts approximately perpendicular (between 30° and 90°) to the weld toe, see Sect. 2.3. If necessary, the other principal stress can be considered using the fatigue class for longitudinally-loaded welds in the nominal stress approach according to Ref. [2].

E. Niemi et al., *Structural Hot-Spot Stress Approach to Fatigue Analysis of Welded Components*, IIW Collection, DOI 10.1007/978-981-10-5568-3_2

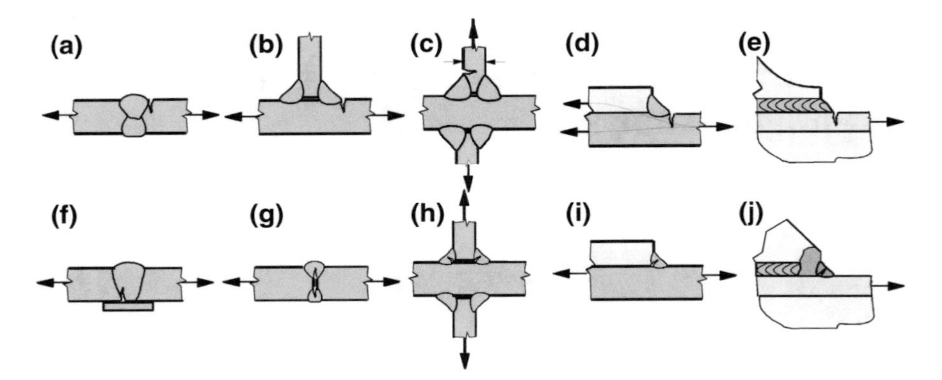

Fig. 2.1 Examples of fatigue crack initiation sites in welded joints

2.2 Types of Hot Spot

Hot spots can be classified as two types, as shown in Fig. 2.2:

Type "a" The weld toe is located on a plate surface, see also Fig. 2.1a–e.
Type "b" The weld toe is located on a plate edge, see also Fig. 2.3.

Different methods are used to determine the hot-spot stress for each type.

Figure 2.3 shows various weld details containing Type "b" hot-spots at the short weld toe or weld end on the plate edge. These welds are classified as load-carrying, except in cases where $L < 100$ mm.

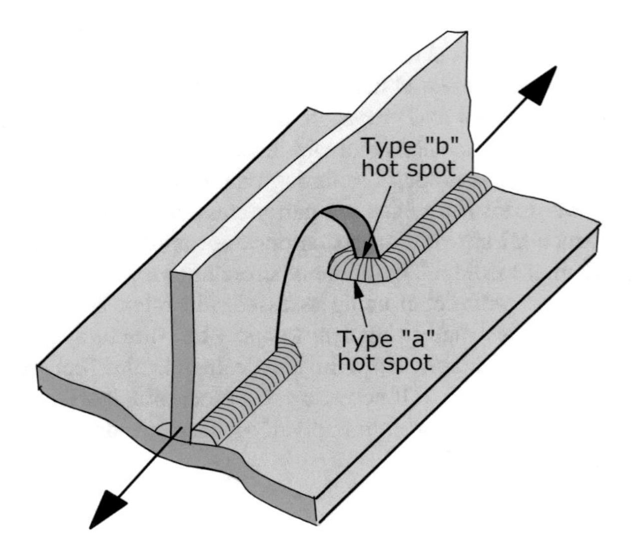

Fig. 2.2 Examples of the two hot-spot types, in a welded girder: Type "a" is located on the surface of the lower flange, Type "b" is located on the edge of the web plate in a scallop

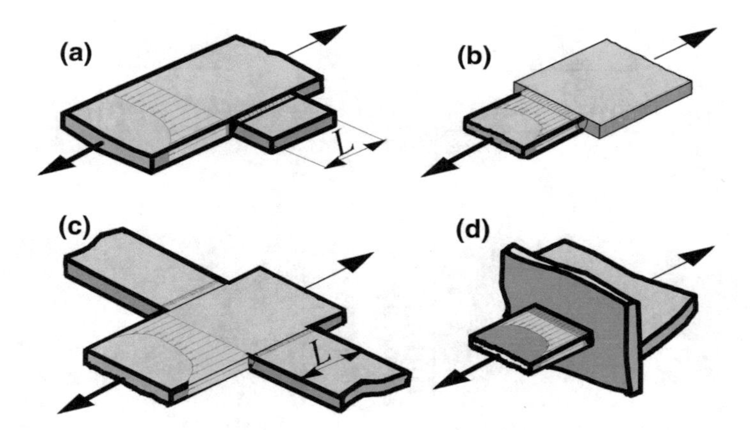

Fig. 2.3 Various details with Type "b" hot-spots at a plate edge, including edge gussets, crossing beam flanges and other details with sharp in-plane notches. *Note* The details shown in Fig. 2.3 represent high structural stress concentrations and correspondingly low fatigue resistance. Therefore, where possible, rounded details and ground corner radii are preferable

In the following sections, recommendations are given concerning mainly Type "a" hot-spots. Type "b" hot-spots are considered in Sects. 3.3 and 4.3.2.

2.3 Definition of the Structural Stress at a Type "a" Hot-Spot

The hot-spot is the critical location at the weld toe (or weld end) where a fatigue crack can be expected to initiate. The structural hot-spot approach is based on the range of the structural stress at the hot-spot (called the "structural hot-spot stress range"). Structural stress, σ_s, is the sum of membrane stress,σ_m, and shell bending stress, σ_b, on the surface of the member, Fig. 2.4.

In many practical cases the larger principal stress range is approximately perpendicular to the weld. Thus, it is interpreted directly as the hot-spot stress. More generally, the hot-spot stress is defined as follows.

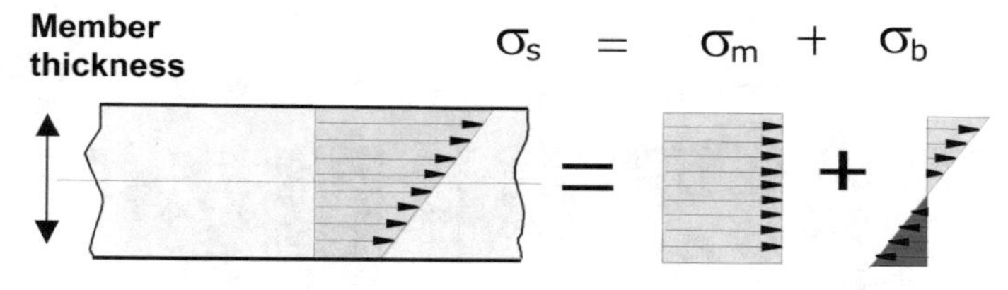

Fig. 2.4 Illustration of the structural stress as the sum of membrane and shell bending stresses on the surface of a member

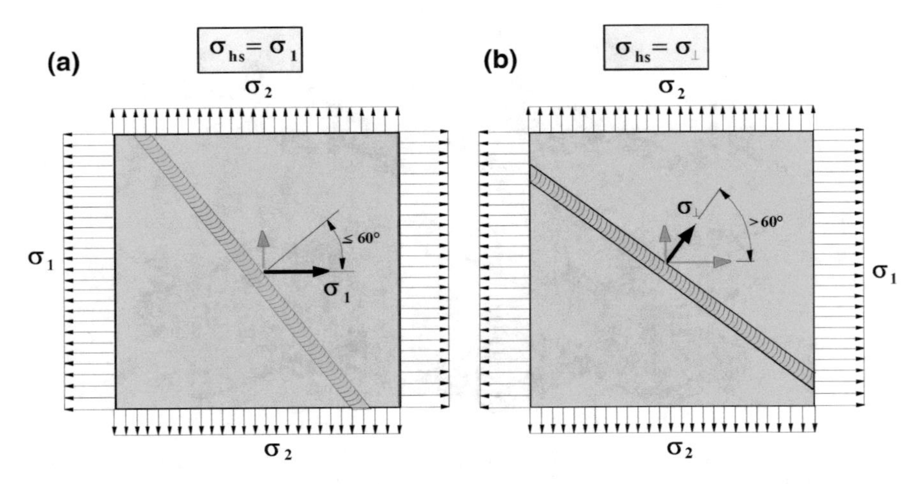

Fig. 2.5 Definition of the stress component used as hot-spot stress in a case when both principal stresses are tensile

- The principal stress with largest range if its direction is within ±60° of the normal to the weld toe, Fig. 2.5a.
- If the direction of the principal stress with largest range is outside the above range, the stress component normal to weld toe, $\sigma_\perp$, Fig. 2.5b, or the minimum principal stress, σ_2, whichever shows the largest range.

In the case of doubt, especially when the directions of the principal stresses change during a load cycle, the partial load factor, γ_F, should be increased sufficiently.

The weld toe represents a local notch, which leads to a non-linear stress distribution through the plate thickness, Fig. 2.6. This consists of three parts: the membrane stress, the shell bending stress and the non-linear stress peak, σ_{nlp}, due to the notch effect of the weld. This notch effect depends on the size and form of the weld and the weld toe geometry. The basic idea of the structural hot-spot approach is to exclude this non-linear stress peak from the structural stress, because the designer cannot know the actual local weld toe geometry in advance. The effect of the notch is implicitly included in the experimentally-determined *S-N* curve. Thus, only the two linearly distributed stress components are included in the structural stress.

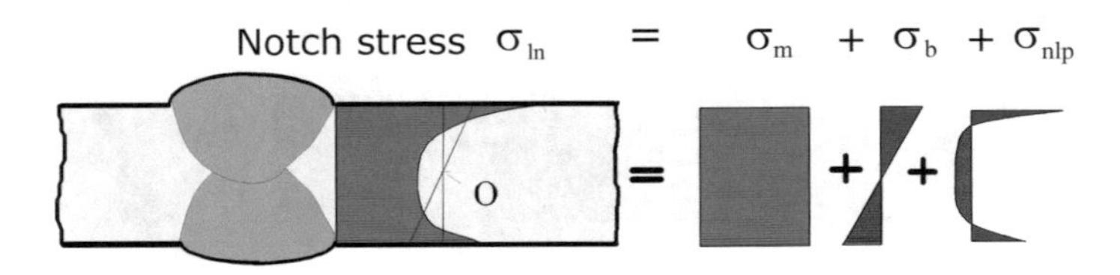

Fig. 2.6 A typical non-linear stress distribution across the plate thickness at a Type "a" hot-spot

It is sometimes argued that the structural hot spot stress is an arbitrary, ill-defined quantity, which does not actually exist. However, according to the above definition, it is an unambiguous quantity, at least for Type "a" hot-spots. Provided the actual non-linear stress distribution is known, the membrane and shell bending stress components can be calculated. However, if it is not known, the structural hot-spot stress must be estimated by extrapolation from the stress distribution approaching the hot-spot.

2.4 Use of Stress Concentration Factors

2.4.1 *Modified Nominal Stress*

The effect of local increases in stress due to geometric features such as structural discontinuities or misalignment can be taken into account by the use of appropriate stress concentration factors. These are used in conjunction with the nominal stress to give the modified nominal stress, σ_{nom}, in the region concerned.

Thus, relevant σ_{nom} must include the effects of the macro-geometric features like large openings, beam curvature, shear lag and eccentricity, as illustrated in Fig. 2.7. Such features are not included in the catalogue of classified details in conventional fatigue design rules based on the use of nominal stress.

2.4.2 *Structural Stress Concentration Factors, K_s*

Stress concentration factors (SCFs) have been published for many types of structural discontinuity. In situations where the nominal stress can be calculated easily, such as weld details in bars or beams, they can be used to estimate the structural hot-spot stress. However, they should be used with care since they might not

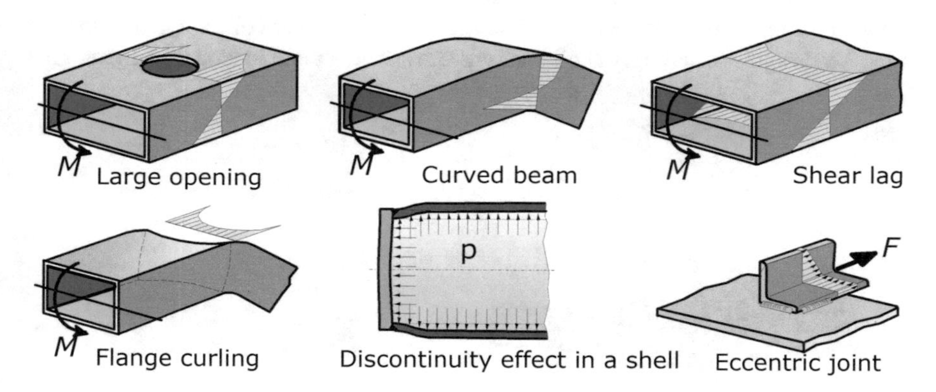

Fig. 2.7 Macro-geometric features which enhance the nominal stress σ_{nom}. In some cases, secondary shell bending stresses are induced

comply with the current definition of structural stress. There is scope for developing valid stress concentration factors, denoted K_s, using FEA and presenting them in the form of parametric formulae. Unfortunately, such formulae are only available in the open literature for a few weld details, see Sect. 5.2.

In principle, the structural hot-spot stress is then given by:

$$\sigma_{hs} = K_s \cdot \sigma_{nom} \tag{2.1a}$$

where σ_{nom} is the modified nominal stress in the area of the hot spot.

In many cases, it is reasonable to use the structural stress concentration factors for axial and bending loading separately. The total structural stress is then:

$$\sigma_{hs} = K_{s,a} \cdot \sigma_{nom,a} + K_{s,b} \cdot \sigma_{nom,b} \tag{2.1b}$$

where

$K_{s,a}$ is the structural stress concentration factor in the case of axial loading
$K_{s,b}$ is the structural stress concentration factor in the case of bending
$\sigma_{nom,a}$ is the modified nominal stress due to the axial loading
$\sigma_{nom,b}$ is the modified nominal stress at the point of interest due to the bending moment

In cases of biaxial bending and even more complicated cases, such as tubular joints with several brace members joined to a chord, the equation can be expanded correspondingly for each hot-spot.

2.4.3 Stress Magnification Factor Due to Misalignment K_m

It is often found that the results of FEA and strain measurements are not in good agreement. One reason is the fact that the FE model is an idealisation of the actual geometry of the structure. In reality, this may be different from the model due to fabrication inaccuracy and welding distortions. Misalignment of the types shown in Fig. 2.8 produces secondary shell bending stresses in a plate loaded by a membrane

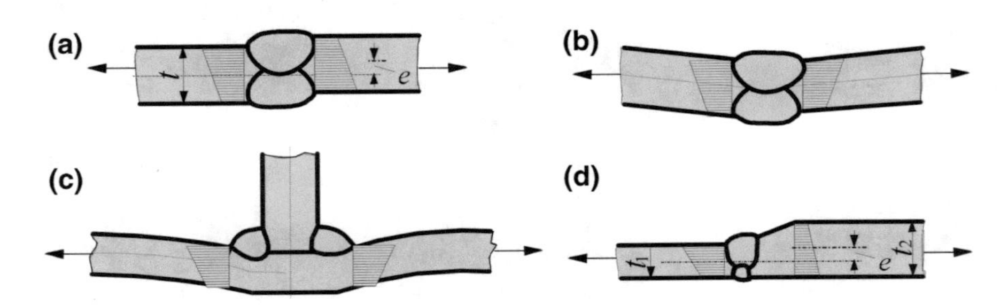

Fig. 2.8 Typical misalignments: **a** offset misalignment, **b** and **c** angular misalignments, **d** eccentric joint

force. Consequently, the designer should always take into account the type and extent of expected misalignment (e.g. based on material or fabrication dimensional tolerances) in the stress calculations.[1] Usually, it is desirable to perform the finite element analysis using a model with ideal geometry, ignoring the fabrication misalignments. In that case, a magnification factor, K_m, can be used for estimation of the modified nominal stress:

$$\sigma_{nom} = K_m \cdot \sigma_{nom,m} + \sigma_{nom,b}, \tag{2.2}$$

where

K_m is a magnification factor, Sect. 5.1;
$\sigma_{nom,m}$ is the membrane part of the nominal stress;
$\sigma_{nom,b}$ is the shell bending part of the nominal stress.

Sometimes the nominal stress components are unknown. Then it is sufficient to replace them in Eq. (2.2) with the structural hot-spot stress components, $\sigma_{hs,m}$ and $\sigma_{hs,b}$. This will be conservative.

It should be noted that in some cases the behaviour of misaligned joints under load is significantly non-linear, depending on the level of applied stress, see Sect. 5.1. In such a case, Eq. (2.2) must be applied for both σ_{max} and σ_{min}. The modified stress range is then the difference between the modified maximum and minimum stresses. However, it is conservative to apply Eq. (2.2) directly to stress range, as would also be appropriate when the K_m factor is independent of the stress level.

It would be advisable to include fabrication tolerances and recommended K_m values for typical details in design guidance developed specifically for particular structures, e.g. cranes and ship hulls.

---

[1]Misalignment is addressed in different ways in the IIW Fatigue Design Recommendations [2], depending on its source:

1. Some misalignment effects are already taken into account in the fatigue classes referred to nominal stresses. The same is true for the structural hot-spot stress classes, but only for the small amount corresponding to up to 5% stress increase [2]. Particularly for butt joints, cruciform joints and transverse attachments, additional misalignment effects resulting from fabrication inaccuracy and welding distortion should be taken into account by increasing the stress or dividing the fatigue class by K_m.
2. Designed eccentricities, as shown in Fig. 2.8d, should be taken into account by calculating the extra shell bending stress.

However, in this Designer's Guide, it is recommended that all sources of extra stress, including misalignments resulting from fabrication inaccuracy and welding distortion, are taken into consideration in the stress calculations.

2.5 Effect of Component Size on the Fatigue Resistance

The fatigue strength of welded joints is size dependent. This size effect is a combination of

- geometric;
- statistical; and
- technological effects.

The geometric effect is often dominant. It depends on the stress gradient in the crack growth direction. In thick plates with geometrical discontinuities like that shown in Fig. 3.1, the non-linear stress peak appears in a relatively deep surface layer (low gradient) which gives rise to faster crack growth and shorter life compared to similar details in thinner plate.

The statistical effect refers to the greater probability of introducing large flaws as the extent of welding increases.

The technological effect refers to other characteristics of welded joints that are influenced by size. For example, the choice of welding process and procedure may be dictated by section thickness and welding residual stresses will be higher in the case of large components.

In the structural hot-spot approach, the geometric size effect is taken into account by multiplying the fatigue strength by a so-called thickness correction factor, $f(t)$, which depends mainly on the thickness of the stressed plate, see Sect. 6.1. For Type "b" hot-spots, the plate thickness has only a small effect on the fatigue strength, because the geometric effect now depends mainly on the width of the plate. At the present time, there is no generally accepted method to take into account the geometrical size effect in such cases. However, the extrapolation method for determining the hot-spot stress from strain gauge measurements presented in Sect. 3.2, with fixed extrapolation points, is also intended to take into account the geometric size effect. This is because the leading gauge picks up traces of the notch stress, which depends on the size of the component.

References

1. Fricke, W.: IIW guideline for the assessment of weld root fatigue. Weld. World **57**, 753–791 (2013)
2. Hobbacher, A.F.: Recommendations for fatigue design of welded joints and components. 2nd edn. Springer (2016)

Chapter 3
Experimental Determination of the Structural Hot-Spot Stress

3.1 General

Experimental stress analysis is usually accomplished on the basis of strain measurements using electric resistance strain gauges. Thus, only information about the stresses on the surface of the component or structure is obtained. In this context, the structural hot-spot stress is established from the stress distribution approaching the weld under consideration, usually on the basis of a particular method of extrapolation.

3.2 Type "a" Hot Spots

Figure 3.1 shows how the stress distribution through the plate thickness changes in the vicinity of a Type "a" hot-spot. At a distance $0.4t$ from the weld toe on the plate surface, the non-linear component has practically vanished and the distribution is almost linear. This fact is exploited in the extrapolation technique used to estimate the structural hot-spot stress, as shown in Fig. 3.2.

In most Type "a" hot-spots, the structural stress and strain increase almost linearly when approaching the weld toe. When the structural hot-spot stress is determined using strain gauges, for example in fatigue testing or the field testing of prototypes, it is sufficient to use linear extrapolation, as shown in Fig. 3.2. Two strain gauges, A and B, are attached $0.4t$ and $1.0t$ from the weld toe, and the structural strain at the hot-spot is determined by linear extrapolation from the strains at these extrapolation points. The same extrapolation points would also be suitable for analyzing stresses obtained by fine mesh FEA, as discussed in Sect. 4.5. In field measurements from prototype testing, these extrapolation points give conservative results when compared to those recommended in Sect. 4.4.3 for use in connection with finite elements of pre-determined size.

E. Niemi et al., *Structural Hot-Spot Stress Approach to Fatigue Analysis of Welded Components*, IIW Collection, DOI 10.1007/978-981-10-5568-3_3

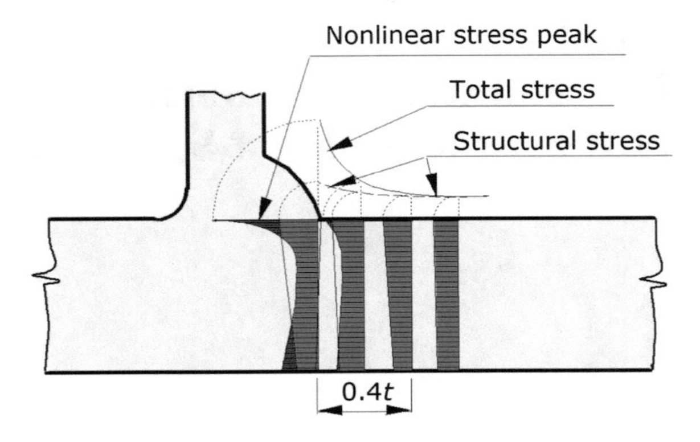

Fig. 3.1 Variation in the through-thickness stress distribution approaching the weld toe

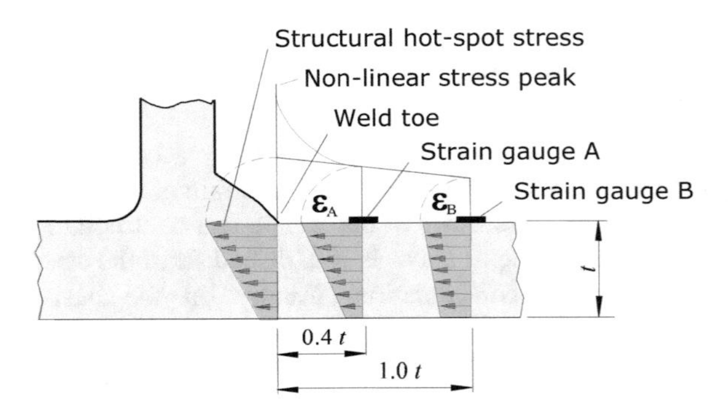

Fig. 3.2 Linear extrapolation to the weld toe in order to estimate the structural hot-spot strain

When the strain gauges are located $0.4t$ and $1.0t$ from the weld toe, the structural hot-spot strain is given by:

$$\varepsilon_{\mathrm{hs}} = 1.67\varepsilon_{\mathrm{A}} - 0.67\varepsilon_{\mathrm{B}}. \tag{3.1}$$

In some cases where the stressed plate is resting on a relatively stiff elastic foundation, such as a beam flange just above a web plate, the stress in the vicinity of a structural discontinuity increases in a non-linear progressive manner as the weld toe is approached. In such cases, linear extrapolation would underestimate the actual structural hot-spot stress. Instead, a quadratic extrapolation method is more suitable. For this, at least three strain gauges, A, B and C, are needed. It is recommended that they are attached at locations $0.4t$, $0.9t$ and $1.4t$ from the weld toe. Then the structural hot spot strain is given by:

$$\varepsilon_{hs} = 2.52\varepsilon_{\mathrm{A}} - 2.24\varepsilon_{\mathrm{B}} + 0.72\varepsilon_{\mathrm{C}}. \tag{3.2}$$

Often multi-element strip strain gauges, with fixed distances between the gauges, are used. Thus, their positions may not coincide exactly with the above extrapolation points. In such circumstances, it is recommended that sufficient gauges are used to enable a curve to be fitted to the results to establish the required strains at the extrapolation points by interpolation.

If the stress state is close to uni-axial, the structural hot-spot stress can be approximated using Eq. (3.3):

$$\sigma_{\text{hs}} = E \cdot \varepsilon_{\text{hs}}. \tag{3.3}$$

However, if the stress state is bi-axial, the actual stress may be up to 10% higher than that obtained from Eq. (3.3). If high accuracy is required, the ratio of the longitudinal and transverse strains, $\varepsilon_{\text{y}}/\varepsilon_{\text{x}}$, should be established from rosette strain gauges or FEA. The structural hot-spot stress, σ_{hs}, can then be calculated (assuming that this principal stress is acting normal to the weld toe) from Eq. (3.4):

$$\sigma_{hs} = E\,\varepsilon_x \cdot \frac{1 + \nu \frac{\varepsilon_y}{\varepsilon_x}}{1 - \nu^2} \tag{3.4}$$

Instead of absolute strains, strain ranges, $\Delta\varepsilon = \varepsilon_{\text{max}} - \varepsilon_{\text{min}}$, are usually measured and substituted in the above equations, producing the structural hot-spot stress range, $\Delta\sigma_{\text{hs}}$.

Again it is emphasised that the extrapolation methods are approximate and sometimes they slightly underestimate the actual structural hot-spot stress. If the hot spot strain being used to present fatigue test results is underestimated, the resulting *S-N* curve will be too low (i.e. conservative). The *S-N* curves given in this document are based on conservative linear extrapolation according to Eq. (3.1). On the other hand, underestimation of the hot-spot stress at the design stage is non-conservative. One source of error is the use of fixed positions for the measuring points for all types of weld detail. Labesse and Récho [1] have shown that more consistent results could be achieved if the measuring points were dependent on the type of the detail, but this will not be discussed further here.

Determination of the structural hot-spot stress would, of course, be much more convenient if it corresponded to the stress at a single point, for example 0.4t from the weld toe. However, details of the stress distribution approaching the weld would also be needed to estimate the stress at the weld toe. In this context, there may be scope for producing stress distributions for a range of typical weld details by FEA. Alternatively, allowance for error could be made simply by increasing the assumed loading by a sufficient partial safety factor, γ_{F}.

3.3 Type "b" Hot Spots

A feature of Type "b" hot spots, which contrasts with Type "a", is that the stress distribution approaching the weld toe does not depend on the plate thickness. Thus, extrapolation points cannot be established as proportions of plate thickness. A tentative method for dealing with this situation was proposed in [2]. This entails the measurement of strains on the plate edge at three absolute distances from the weld toe, or from the weld end if the weld does not continue around the end of the attached plate, namely 4, 8 and 12 mm. After converting strains into stresses, the structural hot-spot stress is then determined by quadratic extrapolation to the weld toe, as follows:

$$\sigma_{hs} = 3\sigma_{4\,mm} - 3\sigma_{8\,mm} + \sigma_{12\,mm} \tag{3.5}$$

Other extrapolation formulae, especially for use with finite element analysis, are given in Sect. 4.

References

1. Labesse, F., Récho, N.: Local–global stress analysis of fillet welded joints. IIW Doc. XIII-1737-98, (1998)
2. Niemi, E.: Stress determination for fatigue analysis of welded components. Abington Publishing, Abington Cambridge (1995)

Chapter 4
Structural Hot-Spot Stress Determination Using Finite Element Analysis

4.1 General

In the design phase, finite element analysis (FEA) is an ideal tool for determining the structural hot-spot stress. It is useful also in production of stress concentration factor formulae for various types of structural detail, see Sect. 5.2.

Linear elastic material behaviour can normally be assumed, since only localized yielding is permitted by most design codes [1]. Since the structural stress range, $\Delta\sigma = \sigma_{max} - \sigma_{min}$, is required, at least two loading cases should normally be analysed, giving the maximum and minimum stresses in the detail in question.

Analysis of large structures with several potential hot-spots can be performed in two phases. First, a coarse model is resolved in order to identify the hot-spot areas. Second, sub-models are created from these areas one at a time, using the nodal displacements or nodal forces from the original model as loading at the boundaries of the sub-model. Another possibility is to refine the original element mesh in the hot-spot regions.

Care is needed to avoid misinterpreting the finite element results, and the following comments should be noted:

- A typical post-processor displays the nodal stress at the weld toe as an average of two elements located on both sides of the weld toe. For post-processing, it is advisable to pick only those elements of interest in front of the weld.
- Results must be obtained that exclude the non-linear stress peak (as shown for Type "a" hot-spots in Fig. 3.1), even in sections close to the weld toe. This will be the case with shell elements, since they automatically exclude this stress peak. If a single-layer solid element mesh is used, a linear distribution is obtained with 8-node elements or with 20-node elements after reduced integration in the thickness direction. If a multi-layer solid element mesh is used, the

E. Niemi et al., *Structural Hot-Spot Stress Approach to Fatigue Analysis of Welded Components*, IIW Collection, DOI 10.1007/978-981-10-5568-3_4

results include a more or less accurate approximation of the non-linear stress distribution. The linearly distributed part, or the structural hot-spot stress, can then be resolved only by dividing the distribution into three parts, as shown in Fig. 2.6.

- In some cases, the weld geometry is neglected in the modelling. Special care is necessary to select an appropriate point to represent the weld toe for stress extrapolation. Frequently, the structural intersection point of the mid-planes or the plate surfaces is chosen because the stress at the weld toe position might be non-conservative.

4.2 Choice of Element Type

The elements used must be able to model plate bending. There are two possibilities:

- shell elements;
- solid elements.

Usually, shell (or plate) elements or one layer of 20-node solid elements yield reasonably accurate results in the assessment of Type "a" hot-spots. It should be noted that the goal is to omit the non-linear stress peak. This implies that 20-node solid elements with reduced 2-point integration in the thickness direction are sufficient. One problem with simple solid elements is the tendency towards the so-called shear locking, which also makes this type of solid element preferable.

If multi-layer modelling of a plate with solid elements is used, which may also be the case when only half the plate thickness is modelled, the surface stress includes part of the non-linear stress peak, up to about $0.4t$ from the weld toe. Therefore, the stress results should be linearized across the thickness, or the stresses should be read outside that area and extrapolated to the weld toe.

Shell elements are used to model the middle planes of plates. Plate thickness is given as a property of the element. Attachments on plates should be extended towards the mid-plane of the plate, or connected to it with rigid links. If the welds are not modelled at all, the stiffness of the plate between adjacent discontinuities becomes too low. For example, in the rectangular hollow section joint shown in Fig. 4.1, the welds should be modelled in order to obtain realistic stiffness in the gap, g', or in the corner. Solid elements are preferable in such details.

Tetrahedron elements are often available, typically in so-called Computer-Aided Engineering (CAE) software, in connection with a feature for easy mesh generation. Linear tetrahedron elements are not suitable for structural stress determination. Only higher order elements, together with a reasonably fine element mesh in the vicinity of the hot-spot, may be used, as long as the user calibrates the method with suitable benchmark data. Usually, a finer mesh is necessary compared to brick elements.

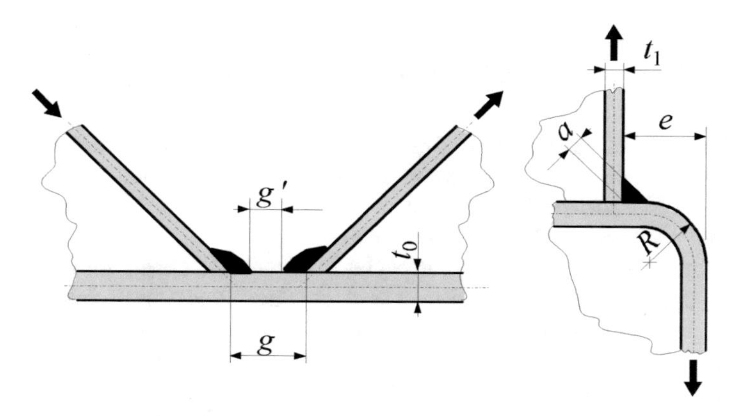

Fig. 4.1 Details in which weld modelling is necessary [2]

4.3 Methods for Determination of Structural Hot-Spot Stress

There are three principal methods for avoiding the influence of the nonlinear stress peak at the weld toe, shown in Fig. 3.2, in the derivation of the structural hot-spot stress σ_{hs}:

1. using the stress distribution linearized through the plate thickness (Fig. 4.2a, b)
2. using the stress distribution extrapolated linearly on the plate surface (Fig. 4.2c) and
3. using a stress at a single point outside the influence area of the nonlinear stress peak, e.g. the intermediate point of the stress distribution in Fig. 4.2c.

The methods are described in the following subsections.

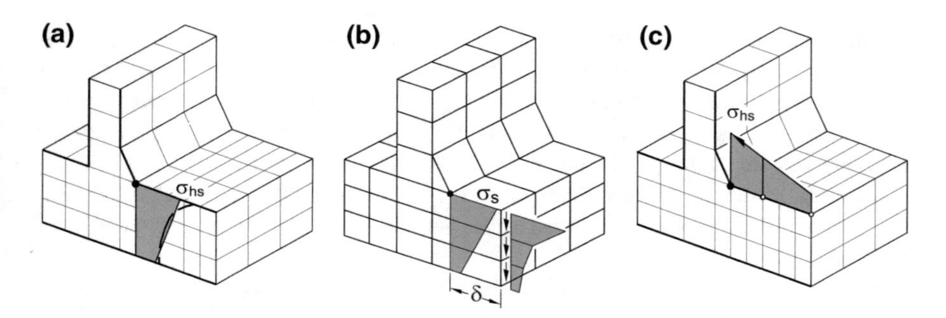

Fig. 4.2 Through-thickness linearization, equilibrium with stresses at distance (**a**) δ and (**b**) surface stress extrapolation to the weld toe (**c**)

4.3.1 Determination of the Structural Stress at the Weld Toe Using Through-Thickness Linearization

Through-thickness linearization Fig. 4.2a is generally applicable to Type “a” hot-spots only. Through-thickness linearization is also possible along the upper weld toe on the surface of the vertical web in Fig. 2.2 up to the corner of the cut-out, i.e. close to the Type “b” hot-spot.

Plate and shell elements naturally yield stresses linearized through the plate thickness, i.e. the sum of membrane and bending stress according to Fig. 2.6. However, in some cases, particularly at attachment ends, compared to solid elements the structural stress might be exaggerated due to the stress singularity, so that surface stress extrapolation is preferred.

With a suitable post-processing program an advanced FE analyst *may be* able to solve the linear stress distribution over the plate thickness directly at the weld toe section in the case of Type “a” hot-spots. This is particularly suitable for meshes with several elements arranged over the plate thickness. Some programs offer a linearization of stress results, which can be applied to the selected row of elements directly in front of the weld toe, see also the remarks given under 4.1 with respect to selecting the elements in front of the weld.

This method is particularly useful for cases where the structural stress distribution in front of the weld toe is nonlinear and for relatively thick structural components. In the latter, location of the read-out points for surface stress extrapolation far away from the weld toe may cause certain geometric effects on the local stress to be missed.

It should be borne in mind that the stress in the element adjacent to the weld toe notch is affected by the stress singularity at the notch so that several elements (at least three) in the thickness direction are required to avoid any influence of the notch and so obtain reasonable results. Dong et al. [3, 4] proposed a special procedure to derive the linear part of the stress distribution in the through-thickness direction. The structural stress is calculated using local stress outputs from solid elements at a certain distance away from the weld toe (Fig. 4.2b) and imposing equilibrium conditions in terms of stress resultants between two adjacent cut sections (Fig. 4.2b). More preferably nodal forces at the weld toe plane can be used to calculate line forces and line moments along the weld toe line, from which membrane and bending stress can be derived. With nodal forces and moments, greater accuracy can be achieved compared to procedures using element stresses.

Dong proposed to linearize the stress in special cases over a certain depth, e.g. the critical crack depth. In this case the method is applicable also to Type “b” hot-spots.

If shell/plate elements are applied, the distribution of the structural stress along the weld toe can be calculated directly from the nodal forces and moments in the elements in front of the weld toe considering the element shape functions. This is done by means of a matrix relating nodal forces/moments to line forces and moments. Then membrane and bending stresses can be calculated at each nodal position along the weld toe line (see [4] and [5]).

Results obtained for different types of weld are compared in [6] and [7] with the alternative method of surface stress extrapolation, which is described in the following section.

4.3.2 Determination of the Structural Stress at the Weld Toe Using Surface Stress Extrapolation

As an alternative to the through-thickness linearization of stresses described above, extrapolation techniques similar to those applied to measured strains (see Sects. 3.2 and 3.3) can be used. In this context, if the FE model of the welded joint considered includes the weld, the extrapolation points are measured from the weld toe. However, if the model consists of shell elements and the weld between the parts joined is not included, it is recommended to measure them from the intersection point of the elements representing those parts in order to avoid non-conservative results (see also 4.4). Also the intersection point of the plate surfaces has been found to be a suitable target for extrapolation in order to avoid over-conservative results [8]. The stresses are obtained at the integration points of adjacent elements, or at the nodal points some distance away from the weld toe. The element mesh must be refined near the hot-spot such that the stress, and the stress gradient, can be determined with an accuracy comparable with that of strain measurements used for experimental hot-spot stress determination, see Sects. 3.1 and 3.2. Typical models, which might be rather coarse, will be described below.

Strictly speaking, the correct method would be first to extrapolate each stress component to the weld toe, and then to resolve the principal stresses and their directions at the hot-spot. However, in practice it is sufficient to extrapolate either the maximum principal stress or the stress component normal to weld toe, whichever is dominant according to Sect. 2.3. In complex details, the directions of the principal stresses tend to change between extrapolation points. This may make it difficult to choose between cases (a) and (b) in Fig. 2.5.

As the stresses in the finite elements are dependent on the mesh density and the element properties, it is necessary to follow some guidelines on the choice of element types and sizes as well on stress evaluation at the extrapolation points. In the FE analysis of large structures (e.g. like ship hulls) it is not practical to use the fine element mesh required for accurate resolution of the stress field in the vicinity of the weld toe. A practical, relatively coarse mesh can then be used if certain conditions are fulfilled, or else a relatively fine mesh must be used. These two meshing strategies are summarized in overview in Table 4.1 and Fig. 4.3, and explained in more detail in Sects. 4.4 and 4.5.

Sometimes there are two adjacent welds so close together, as in Fig. 4.1, that neither of the above-mentioned meshing and extrapolation strategies is possible.

Table 4.1 Guideline on meshing and stress evaluation using surface stress extrapolation (further details are described in Sects. 4.4 and 4.5)

Types of model and weld toe:		Relatively coarse models		Relatively fine models	
		Type a	Type b	Type a	Type b
Element Length × Width	Shells:	$\leq t \times t$ and $t \times w/2$[a]	10 mm × 10 mm	$\leq 0.4t \times t$ and $0.4t \times w/2$[a]	≤4 mm × 4 mm
	Solids:	$\leq t \times t$ and $t \times w$[a]	10 mm × 10 mm	$\leq 0.4t \times t$ and $0.4t \times w/2$[a]	≤4 mm × 4 mm
Extrapol. Points	Shells:	$0.5t/1.5t$ (mid-side pts.)[b]	5 mm/15 mm (mid-side points)	$0.4t/1.0t$ (nodal points)	4 mm/8 mm/12 mm (nodal points)
	Solids:	$0.5t/1.5t$ (surface centre)	5 mm/15 mm (surface centre)	$0.4t/1.0t$ (nodal points)	4 mm/8 mm/12 mm (nodal points)

[a]w = attachment width (attachment thickness +2 weld leg lengths), see also 4.4

[b]surface centre at transverse welds, if the weld below the plate is not modelled, see Fig. 4.4c, d

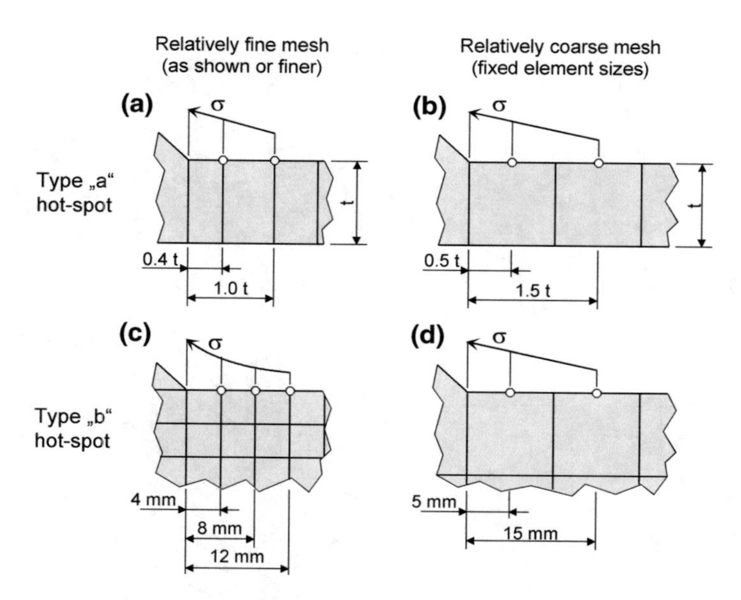

Fig. 4.3 Guideline on meshing and stress evaluation using surface stress extrapolation (further details are described in Sects. 4.4, 4.5 and in Table 4.1)

Then a sufficiently fine mesh should be designed, so that the structural hot-spot stress can be determined by through-thickness linearization of stresses or else by curve fitting using stresses from at least three points.

The mesh pattern for Type "a" hot spots should be designed so that the actual hot-spot can be found, and the structural stress can be determined by extrapolation at this location. Symmetry lines are not always the correct paths. Extra care is needed in complex details, such as tubular joints. In a plate subject to uniaxial membrane stress, the location of the hot-spot may be obvious. However, where the loading introduces plate bending stresses and the weld is curved around attachment corners, the location of the hot-spot may be less obvious. Then the mesh pattern should allow the determination of structural stresses at all potential locations. For example, the mesh patterns shown in Figs. 4.5 and 4.6 are suitable mainly for longitudinal membrane loading. If some loads were acting on the bracket causing bending of the main plate, a mesh pattern resembling a polar co-ordinate system around the stiffener end should be designed, as shown in Fig. 7.3.

In case of Type "b" hot-spots in plate edge attachment details, both the distribution of the structural stress caused by the gusset geometry and that due to the effect of the weld fillet are non-linear and occur in the same plane. Therefore, it is not easy to distinguish between the local effect of weld toe geometry and the effect of the structural discontinuity. In future, theoretically more correct methods for doing this may be available. Meanwhile, two tentative methods have been shown to yield consistent results [9], see Fig. 4.3c, d.

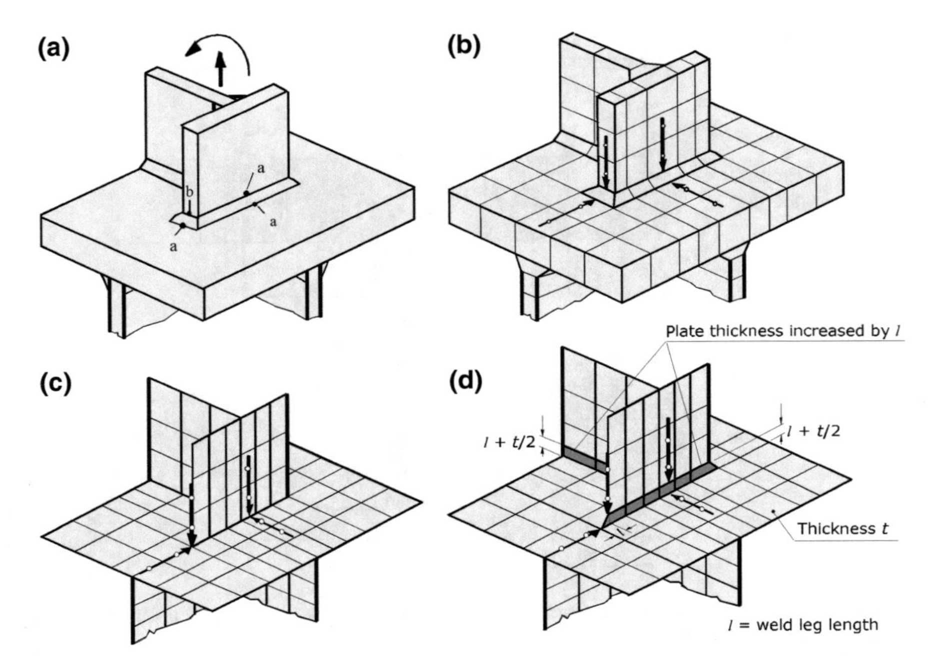

Fig. 4.4 Examples of relative coarse meshing, including stress evaluation points and extrapolation paths: **a** welded component showing potential hot-spots of Types "a" and "b"; **b** solid element model with extrapolation points in the centre points of the elements; **c** shell element model with extrapolation points for the weld ends located at the mid-side nodes and for the transverse welds at the centre points of the elements; **d** the welds of the attachments modelled with reinforced plate strips with inclined ends

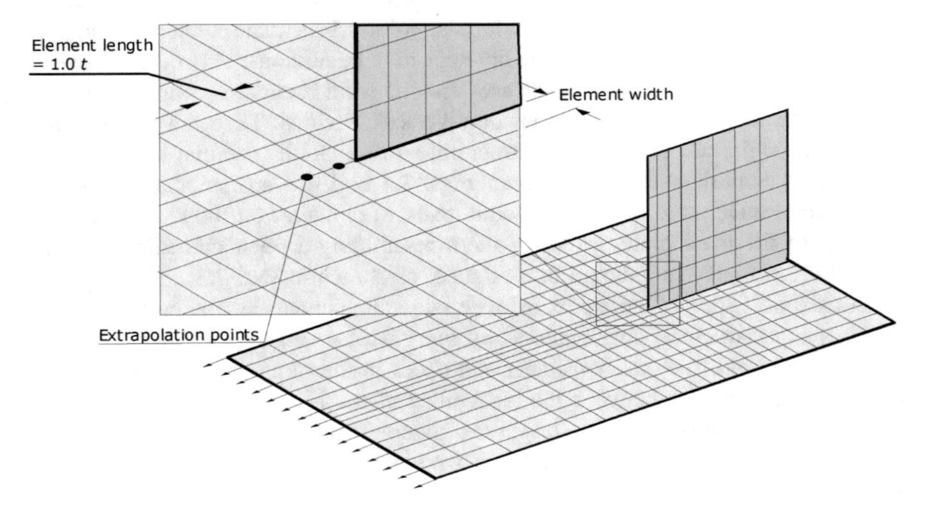

Fig. 4.5 An example of a local coarse shell model at a longitudinal gusset

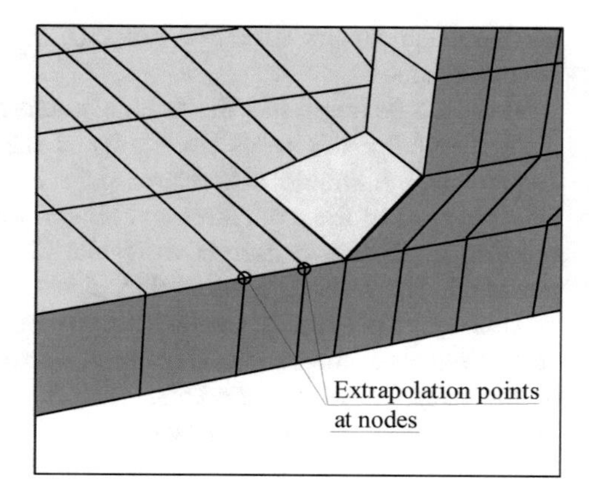

Fig. 4.6 An example of simple solid element modelling of a bracket welded on a plate. The darkest face is a symmetry plane

The first method, used with relatively fine meshes, corresponds to the experimental method described in 3.3, where strain gauges are placed at distances 4, 8 and 12 mm from the weld toe. The second method is more practical using a relatively coarse element mesh, based on Refs. [9, 10]. In this method, the singularity at the weld toe exaggerates the mid-point stress in the first element such that linear extrapolation yields results comparable with those obtained with more refined analysis and extrapolation according to Eq. (3.5).

4.3.3 Determination of the Structural Stress at a Single Point Close to the Weld Toe

Proposals have been made to simplify the structural hot-spot stress evaluation by using the surface stress outside the area influenced by the weld toe notch, e.g. $0.5 \times$ plate thickness t from the toe [11, 12]. The mesh density should be chosen accordingly. If plate or shell elements are used without modelling the weld, the location of the stress read-out point should refer to the intersection points of the plates (or at least the plate surfaces) as explained in the previous section.

As the structural stress distribution usually contains a stress gradient, the structural hot-spot stress at a distance of $0.5t$ is smaller than that at the weld toe. Therefore, it is used with a lower design *S-N* curve, i.e. a lower FAT class, see also Sect. 6.2.1.

Alternative proposals using absolute distances from the weld toe have been made by Haibach and later by Xiao and Yamada. Haibach [13] proposed to utilize the strain on the plate surface 2 mm away from the weld toe, finding from measurements a common strain-life curve for various welded details having different geometry. Xiao and Yamada [14] proposed to compute the stress at a depth of 1 mm below the weld toe on the potential crack path. They showed good correlation between this stress and fatigue life to the extent that it could be considered to be equivalent to the structural hot-spot stress and used with the same design *S*-*N* curves.

It should be noted that the finite element mesh has to be fine enough to exclude the influence of the stress singularity at the weld toe on the read-out point. The element length should not exceed 0.5 mm. Furthermore, the maximum principal stress instead of the axial stress parallel to the plate surface should be evaluated in order to avoid non-conservative results [15]. The approach has been found to be unsuitable for welded joints in thin plates ($t \leq 5$ mm) [16].

Haibach's and Xiao/Yamada's approaches consider the plate thickness effect to a certain extent as the structural stress at a fixed distance from the weld toe increases as the plate thickness increases. Therefore, no additional thickness correction is necessary with these approaches.

4.4 Use of Relatively Coarse Element Meshes

Several classification societies [17–20] have developed practical extrapolation techniques that allow the use of a relatively coarse element mesh. These have been further investigated [2, 6, 8, 9, 17] leading to the recommendations in this guide (see also Figs. 4.3b, 4.4 and Table 4.1). Apart from the direct fatigue assessment, relatively coarse meshes can be used for screening structures to find fatigue-critical locations.

Use of a relatively coarse element mesh is sufficient when the following conditions are satisfied:

- 8-node shell elements or 20-node solid elements are in use (alternatively 4-node shell elements with additional degrees of freedom may be used);
- stresses are resolved at mid-points or mid-side nodes of the elements;
- the structural hot-spot stress may be determined by extrapolating surface stresses to the hot spot or taking the stress at a single point $0.5 \times t$ away from the hot spot.

In this method, the stress obtained at mid points or mid-side nodes of the first elements is slightly exaggerated due to the singularity at the weld toe. This compensates for the error caused by linear extrapolation from relatively distant extrapolation points.

4.4.1 Solid Element Modelling

Solid element modelling allows the weld geometry to be easily included, Fig. 4.4b, which is particularly important in complex details. Iso-parametric 20-node solid elements with reduced integration should be used. Only one layer of elements is required over the whole plate thickness.

Solid element lengths at hot spots of type (a), in the direction of the stress, should be no greater than $1.0t$. The element width can usually be taken as the width of the attachment, defined as attachment thickness plus twice the weld leg length, (see Table 4.1), but not more than the plate thickness. However, very wide attachments may require smaller elements.

At Type "b" hot spots, fixed element lengths should be chosen according to Fig. 4.3d.

4.4.2 Thin Shell (or Plate) Element Modelling

Thin plate or shell elements have to be arranged at the mid-planes of the structural parts, Fig. 4.4c. Eight-node elements are recommended, particularly in cases of steep stress gradients. The welds are not usually modelled. However, in cases of high local bending between two local discontinuities (see Fig. 4.1), welds could be modelled by reinforced plate strips (see Fig. 4.4d) or by the methods described in Sect. 4.6.

The shell element length in the direction of the stress should normally be $1.0t$ at Type "a" hot spots, as shown in Fig. 4.5. At the ends of longitudinal brackets and similar details, the results may be rather sensitive to the width of the element adjacent to the attachment. Elements that are too wide can give non-conservative results. In typical details, the width should be equal to the main plate thickness but no greater than half of the attachment width, see Table 4.1. At Type "b" hot spots, the fixed lengths given in Table 4.1 apply.

4.4.3 Hot-Spot Stress Extrapolation

The linear extrapolation at Type "a" hot spots is performed using stresses $\sigma_{0,5t}$ and $\sigma_{1,5t}$ located at distances $0.5t$ and $1.5t$ from the weld toe respectively, see Fig. 4.3b and 4.4b–d:

$$\sigma_{hs} = 1.5\sigma_{0.5t} - 0.5\sigma_{1.5t} \tag{4.1}$$

For Type "b" hot spots, the same formula can be used taking the stresses σ_1 at 5 mm and σ_2 at 15 mm distance from the hot spot:

$$\sigma_{hs} = 1.5\sigma_1 - 0.5\sigma_2 \tag{4.2}$$

Usually, only the mid-plane stresses are considered because plate bending is taken into account in the structural hot-spot stress on the plate surface.

The locations of extrapolation points are measured from the weld toe modelled (solid elements) or from the intersection of shells (shell element modelling without welds). As mentioned above, the intersection point of the plate surfaces may also be suitable.

This method presupposes that the stresses are read at mid-points on the surface of solid elements, or at the mid-side nodes of 8-node shell elements in connection with the element size specified. One exception concerns Type "a" hot spots above a web plate, as shown in the extrapolation paths on the right in Fig. 4.4(c, d). As the stress extrapolation in shell models tends to exaggerate the stress compared to solid models, extrapolation may be performed from element mid-points. The extrapolation formula (4.1) should not be used in connection with finer element meshes or strain gauge measurements because it could underestimate the hot-spot stress.

4.5 Use of Relatively Fine Element Meshes

Use of a relatively fine element mesh is recommended for the following:

- in complex details and details showing a high stress gradient close to the hot-spot;
- when evaluating the structural hot-spot stress by through-thickness linearization or at a single point close to the hot-spot (surface stress extrapolation is possible as well);
- when comparing FEA results and measured stresses.

4.5.1 Solid Element Modelling

Solid element modelling allows the weld geometry to be considered, as illustrated in Fig. 4.2. Since the extrapolation method uses nodal stresses $0.4t$ and $1.0t$ from the weld toe when applied to Type "a" hot spots, the first element length should be no more than $0.4t$ and the second element length no more than $0.6t$ in the loading direction, Fig. 4.6. For Type "b" hot spots, the required maximum element sizes are shown in Fig. 4.3c. Also the element width should be smaller than in coarse models, see Table 4.1. If a refined mesh is used, it is generally recommended to subdivide the mesh shown in Fig. 4.3(a, c) in both directions, i.e. in the direction of stress extrapolation as well as in the thickness direction.

Refined meshes are generally necessary for evaluation of the structural hot-spot stress by through-thickness linearization or at a single point close to the hot spot, see Sects. 4.3.1 and 4.3.3.

4.5.2 Thin Shell (or Plate) Element Modelling

Thin shell or plate modelling is also possible with finer meshes. The weld modelling recommendations in Sect. 4.4.2 apply. Stress evaluation at a single point close to the hot-spot is, however, no longer possible.

The maximum element sizes are summarized in Table 4.1. Special attention should be paid to longitudinal attachments welded to a flange or plate, as illustrated in Fig. 4.4. When such a detail is modelled with a fine thin shell element mesh without modelling the fillet welds, the entire attachment is connected to only one nodal line. In the vicinity of the attachment, the membrane stress concentration seems to be significantly exaggerated. In reality, the width of the attachment is equal to the thickness of the bracket plus twice the leg length of the fillet weld. A realistic solid element model gives much lower stress concentrations. Therefore, the results should be corrected by suitably averaging the stress within the width of the attachment [2].

4.5.3 Hot-Spot Stress Extrapolation

For Type "a" hot spots, the linear extrapolation procedure presented earlier for the analysis of measured strains is recommended (see Figs. 3.2 and 4.3a):

$$\sigma_{\mathrm{hs}} = 1.67\sigma_{0.4t} - 0.67\sigma_{1.0t} \tag{4.3}$$

where $\sigma_{0.4t}$ and $\sigma_{1.0t}$ are stresses at distances 0.4 t and 1.0 t from the weld toe.

For Type "b" hot spots, Eq. (3.5) applies. The other recommendations in 4.4.3 are also valid here.

4.6 Modelling Fillet Welds in Shell Element Models

In the refined parts of shell element models, it is sometimes important to model the welds as well. This can be carried out using inclined elements, as shown for example in Fig. 4.7a. This model exaggerates the cross-sectional area close to the intersection, which is a drawback in longitudinally loaded joints.

Another possibility is to use inclined elements only in the fillet welds and additional rigid links to the elements in the mid-plane of the plates, see Fig. 4.7b. In this way, the cross sectional area of the elements corresponds to the actual area, and also the stiffness is better idealized. The connection of the rigid links to the plates might require the application of multi-point constraints as connection to the actual nodes [21].

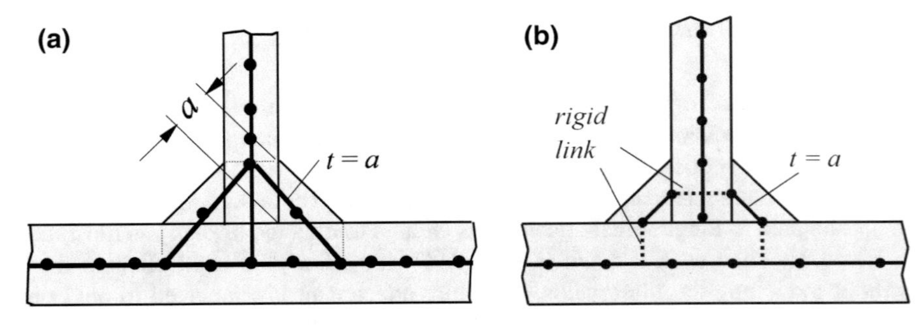

Fig. 4.7 Fillet welds in a T-joint modelled with inclined elements having **a** mid-side nodes or **b** connections with rigid links

References

1. Hobbacher, A.F.: Recommendations for fatigue design of welded joints and components. 2nd Edition, Springer (2016)
2. Niemi, E.: Stress determination for fatigue analysis of welded components. Abington Publishing, Abington Cambridge (1995)
3. Dong, P.: A structural stress definition and numerical implementation for fatigue analysis of welded joints. Intl. J Fatigue **23**, 865–876 (2001)
4. Dong, P., et al.: The master S-N curve method: an implementation for fatigue evaluation of welded components in the 2007 ASME B&PV Code, Section VIII, Division 2 and API 579-1/ASME FFS-1, WRC Bulletin, No. 523, New York (2010)
5. Efthymiou, M: Development of SCF formulae and generalised influence functions for use in fatigue analysis. Proceedings of the Offshore Tubular Joints Conference (OTJ '88), Anugraha Centre, Egham, UK (1988)
6. Doerk, O., Fricke, W., Weissenborn, C.: Comparison of different calculation methods for structural stresses at welded joints. Int. J. of Fatigue **25**, 359–369 (2003)
7. Poutianen, I., et al.: Determination of the structural hot spot stress using the finite element method - a comparison of current procedures. IIW-Doc. XIII-1991-03 and XV-1148-03, (2003)
8. Osawa, N., Yamamoto, N., Fukuoka, T., Sawamura, J., Nagai, H., Maeda, S.: Study on the preciseness of hot spot stress of web-stiffened cruciform welded joints derived from shell finite element analyses. Mar. Struct. **24**(3), 207–238 (2011)
9. Fricke, W., Bodgan, R: Determination of hot spot stress in structural members with in-plane notches using a coarse element mesh. IIW Doc XIII-1870-01, (2001)
10. Wagner M: Fatigue strength of structural members with in-plane notches. IIW Doc. XIII-1730-98, (1998)
11. Maddox, S.J.: Hot-spot stress design curves for fatigue assessment of welded structures. Intl. J Offshore and Polar Engng. **12**(2), 134–141 (2002)
12. Lotsberg, I., Sigurdsson, G.: Hot-spot *S-N* curve for fatigue analysis of plated structures. ASME, Proc. of OMAE Specialty Symp. on FPSO Integrity (2004)
13. Haibach, E.: Die Schwingfestigkeit von Schweissverbindungen aus der Sicht einer örtlichen Beanspruchungsmessung (The fatigue strength of welded joints considered on the basis of a local stress measurement), LBF-Bericht FB-77. Fraunhofer-Inst. für Betriebsfestigkeit, Darmstadt (1986)
14. Xiao, Z.G., Yamada, K.: A method of determining geometric stress for fatigue strength evaluation of steel welded joints. Int. J. Fatigue **26**, 1277–1293 (2004)

15. Fricke, W., Feltz, O.: Fatigue tests and numerical analyses of partial-load and full-load carrying fillet welds at cover plates and lap joints. Weld. World **54**(7/8), R225–R233 (2010)
16. Remes, H., Fricke, W.: Influencing factors on fatigue strength of welded thin plates based on structural stress assessment. Weld. World **58**(6), 915–923 (2014)
17. Fricke, W: Recommended hot spot analysis procedure for structural details of ships and FPSOs based on round-robin FE analyses. Int. J. of Offshore and Polar Engng. 12, No. 1, pp. 40–47 (2002)
18. GL: Rules for Classification and Construction, Part I, Ship Technology, 1.1 Seagoing ships–Hull, Part V, Analysis Techniques, Edition 1998. Germanischer Lloyd, Hamburg (1998)
19. BV: Fatigue Strength of Welded Ship Structures. Publication NI 393 DSM R01 E. Bureau Veritas, Paris (1998)
20. DNV: Fatigue Assessment of Ship Structures. Classification Notes No. 30.7. Det Norske Veritas, Høvik, September (1998)
21. Turlier, D., Klein, P., Bérard, F: "Seam Sim" method for seam weld structural assessment within a global structure FEA. Proc. Int. Conf. IIW2010 Istanbul (Turkey). AWST 651-658 (2010)

Chapter 5
Parametric Formulae

5.1 Misalignment

For the time being, available stress magnification factor (K_m) values are based mainly on 2-dimensional stress analysis. In reality, plated structures are 3-dimensional, often containing longitudinal and transverse stiffeners [1]. Adjacent stiffeners affect the boundary conditions but this is taken into account in a simplified manner in the following formulae [2, 3].

Note that the secondary bending stress due to misalignment depends on the nature of the applied loading. If this is purely shell bending, then misalignment does not introduce any secondary bending at all and $K_m = 1.0$. Similarly, for combined axial and bending loading, K_m applies only to the axial stress component.

The following sub-sections present solutions for estimating K_m for various types of misalignment. They refer specifically to misaligned joints under axial loading; all the sketches indicate tensile loading. In each case, both tensile and compressive secondary bending stresses are induced under either tensile or compressive axial loading. Thus, the resulting total stress in some locations decreases as a result of the secondary bending stress induced, while in others it increases. All the solutions presented refer to the increase in applied stress due to misalignment.

5.1.1 Axial Misalignment Between Flat Plates of Equal Thickness Under Axial Loading

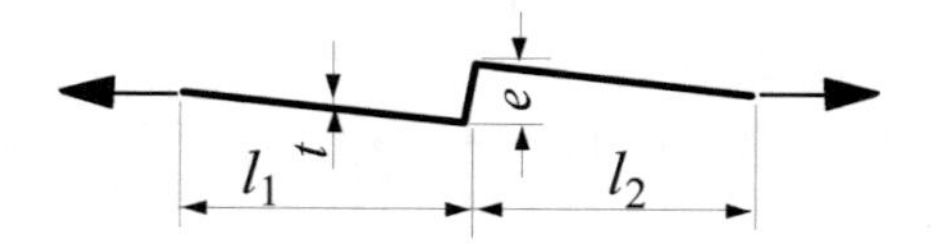

E. Niemi et al., *Structural Hot-Spot Stress Approach to Fatigue Analysis of Welded Components*, IIW Collection, DOI 10.1007/978-981-10-5568-3_5

$$K_m = 1 + \lambda \cdot \frac{e \cdot l_1}{t(l_1 + l_2)} \tag{5.1}$$

λ is dependent on restraint,
$\lambda = 6$ for unrestrained joints.
For remotely loaded joints assume $l_1 = l_2$.

5.1.2 *Axial Misalignment Between Flat Plates of Differing Thickness Under Axial Loading*

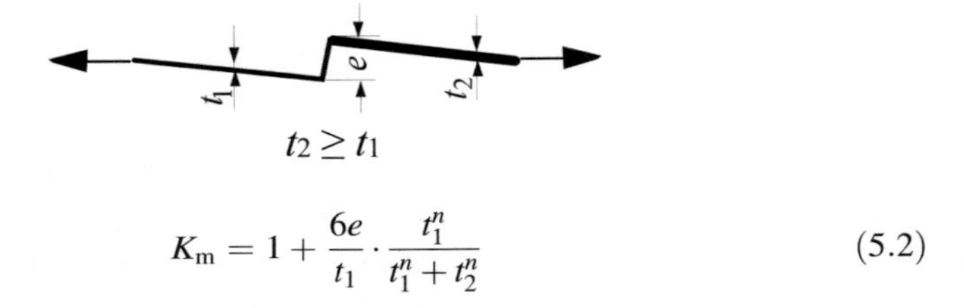

$$K_m = 1 + \frac{6e}{t_1} \cdot \frac{t_1^n}{t_1^n + t_2^n} \tag{5.2}$$

Relates to remotely loaded unrestrained joints.
The use of $n = 1.5$ is supported by tests.

5.1.3 *Axial Misalignment Between Tubes or Pipes Under Axial Loading*

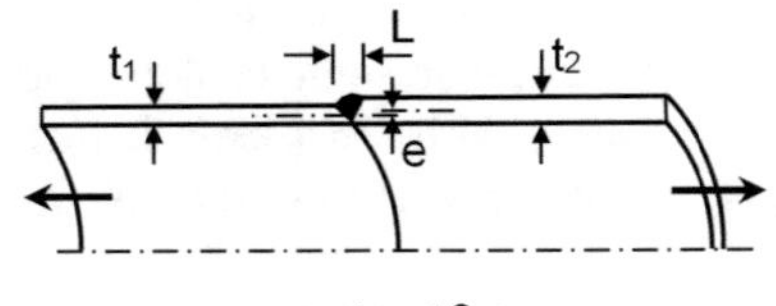

$t_1 \leq t_2 \leq 2\, t_1$

$$K_m = 1 + 6\,\frac{e}{t_1}\left[\frac{1}{1 + \left(\frac{t_2}{t_1}\right)^{\beta}}\right]\exp(-\alpha) \tag{5.3}$$

where $\alpha = \frac{1.82\,L}{\sqrt{Dt_1}} \cdot \frac{1}{1 + \left(\frac{t_2}{t_1}\right)^{\beta}}$ and $\beta = 1.5 - \frac{1.0}{\log\left(\frac{D}{t_1}\right)} + \frac{3.0}{\left[\log\left(\frac{D}{t_1}\right)\right]^2}$ acc. to [4].

L = length over which the eccentricity is distributed,
D = outside diameter of tube or pipe.

5.1.4 Axial Misalignment at Joints in Pressurized Cylindrical Shells with Thickness Change

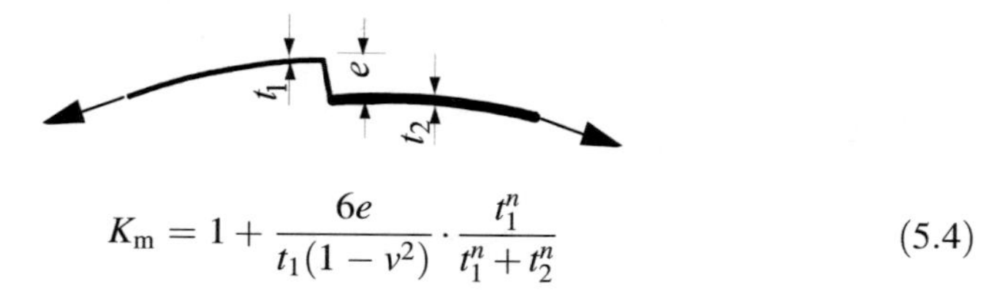

$$K_m = 1 + \frac{6e}{t_1(1-\nu^2)} \cdot \frac{t_1^n}{t_1^n + t_2^n} \tag{5.4}$$

$n = 1.5$ in circumferential joints and joints in spheres.
$n = 0.6$ for longitudinal joints.

5.1.5 Angular Misalignment Between Flat Plates of Equal Thickness Under Axial Loading

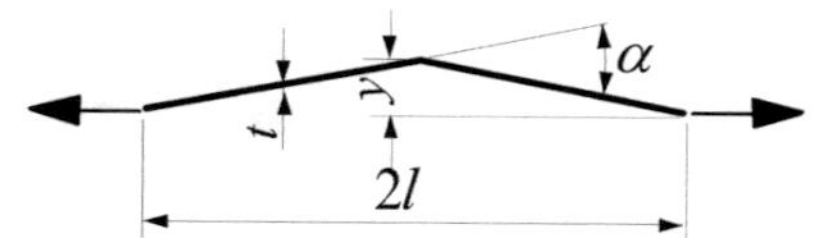

The tanhβ/β etc. correction allows for the reduction of angular misalignment due to the straightening of the joint under tensile loading. It is always ≤ 1 under tension and it is therefore conservative to ignore it. However, in the absence of lateral constraint, angular misalignment increases as an applied compressive load is increased. In such circumstances, the 'tanh' term becomes 'tan' and it is no longer conservative to ignore it.

(i) fixed ends:

$$K_m = 1 + \frac{3y}{t} \cdot \frac{tanh(\beta/2)}{\beta/2} \tag{5.5}$$

where $\beta = \frac{2l}{t}\sqrt{\frac{3\,\sigma_m}{E}}$,

σ_m is membrane stress,
E is modulus of elasticity.

or alternatively:

$$K_m = 1 + \frac{3}{2} \cdot \frac{\alpha \cdot l}{t} \cdot \frac{tanh(\beta/2)}{\beta/2}, \tag{5.6}$$

(ii) pinned ends:

$$K_{\mathrm{m}} = 1 + \frac{6y}{t} \cdot \frac{\tanh(\beta)}{\beta} \tag{5.7}$$

or alternatively:

$$K_{\mathrm{m}} = 1 + \frac{3\alpha \cdot l}{t} \cdot \frac{tanh(\beta)}{\beta} \tag{5.8}$$

5.1.6 *Angular Misalignment at Longitudinal Joints in Pressurized Cylindrical Shells*

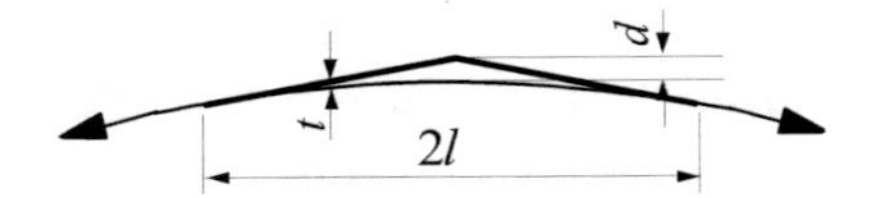

(i) fixed ends:

$$K_{\mathrm{m}} = 1 + \frac{3d}{t(1 - \nu^2)} \cdot \frac{tanh(\beta/2)}{\beta/2} \tag{5.9}$$

$$\beta = \frac{2l}{t}\sqrt{\frac{3(1 - \nu^2) \cdot \sigma}{E}} \tag{5.10}$$

(ii) pinned ends:

$$K_{\mathrm{m}} = 1 + \frac{6d}{t(1 - \nu^2)} \cdot \frac{\tanh(\beta)}{\beta} \tag{5.11}$$

5.1.7 Ovality in Pressurized Cylindrical Pipes and Shells

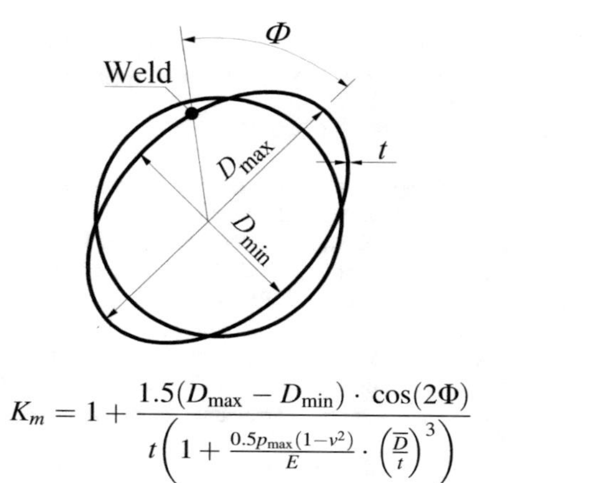

$$K_m = 1 + \frac{1.5(D_{\max} - D_{\min}) \cdot \cos(2\Phi)}{t\left(1 + \frac{0.5p_{\max}(1-\nu^2)}{E} \cdot \left(\frac{\overline{D}}{t}\right)^3\right)} \quad (5.12)$$

where $\overline{D}$ is the average diameter.

If $p_{\max}$ varies under fatigue loading, the mean value may be used during the time interval considered according to [5].

5.2 Structural Discontinuities

Only limited parametric formulae for K_s have been published; they are referred to in Table 5.1.

Table 5.1 Sources of structural stress concentration factors

Application field	Reference
Ship hulls	[6–8]
Pressure vessels	[9]
Reinforced openings	[10]
Tubular joints	[11, 12]

References

1. Partanen, T.: Factors affecting the fatigue behaviour of misaligned transverse butt joints in stiffened plate structures, In: M. Bramat (Ed): Engineering Design in Welded Constructions, Proc. IIW Conf. Madrid (1992)
2. Hobbacher, A.F.: Recommendations for fatigue design of welded joints and components. 2nd edn, Springer (2016)
3. Maddox, S.J.: Fitness-for-purpose assessment of misalignment in transverse butt welds subject to fatigue loading. Welding Institute Research Report 279/1985 Cambridge (UK)
4. Lotsberg, I.: Stress concentrations due to misalignments at butt welds in plated structures and at girth welds in tubulars. Int. J. Fatigue **31**, 1337–1345 (2009)
5. B.S. 7910:2013, Assessing the acceptability of flaws in metallic structures, British Standards Institution, London (2013)
6. G.L.: Rules for classification and construction, Part I, Ship technology, 1.1 Seagoing ships—Hull, Part V, Analysis Techniques, edn 1998. Germanischer Lloyd, Hamburg (1998)
7. B.V.: Fatigue strength of welded ship structures. Publication NI 393 DSM R01 E. Bureau Veritas, Paris (1998)
8. D.N.V.: Fatigue assessment of ship structures. Classification notes No. 30.7. Det Norske Veritas, Høvik (1998)
9. Decock, J.: Determination of stress concentration factors and fatigue assessment of flush and extruded nozzles in welded pressure vessels. 2nd international conference on pressure vessel technology, Part II, ASME, Paper II-59, 821–834 October 1973
10. NORSOK.: Standard N-004, Design of steel structures, Annex C, Fatigue strength analysis, Rev. 1 (1998)
11. Zhao, X.-L., Packer, J.A.: Fatigue design procedure for welded hollow section joints, Recommendations of IIW Sub-commission XV-E. Woodhead Publishing, Cambridge (2000)
12. Efthymiou, M.: Development of SCF formulae and generalised influence functions for use in fatigue analysis. In: Proceedings of the offshore tubular joints conference (OTJ '88), Anugraha Centre, Egham, UK 1988

Chapter 6
Structural Hot-Spot *S-N* Curves

6.1 General Principles

The recommended structural hot-spot stress design *S-N* curves are expressed in the same form as the usual IIW design curves [1], by reference to a fatigue class, *FAT* number, corresponding to the allowable stress range in N/mm^2 or MPa for a fatigue life of 2×10^6 cycles. However, in this case the *FAT* number refers to the hot-spot stress range $\Delta\sigma_{hs}$ rather than the nominal applied stress range. The general form of the *S-N* curve is illustrated in Fig. 6.1. Its equation is:

$$\Delta\sigma_{hs}^{m} \cdot N = C \tag{6.1}$$

where the following notation applies:

FAT	is fatigue class, i.e. the fatigue strength at 2×10^6 cycles;
$\Delta\sigma_{hs}$	is the structural hot-spot stress range;
N	is number of cycles to failure;
$\Delta\sigma_{R,L}$	is constant amplitude fatigue limit, assumed to occur at $N = 10^7$ cycles in [1];
m	is the slope of the upper part of the *S-N* curve (usually $m = 3$ for welds);
C	is design value of "fatigue capacity" ($= 2 \times 10^6 \cdot FAT^m$);

The design curves represent lower bound fatigue strength in non-corrosive environmental conditions, based on statistical analysis of relevant fatigue test data. In general, they are close to the mean minus two standard deviations of log *N* and therefore correspond to approximately 2.3% probability of failure. The effect of the presence of high tensile residual stresses, as can be expected to arise in actual welded structures, has been taken into account.

In the case of post-weld heat-treated or other components known to contain less severe residual stresses, a bonus factor may be applied to the fatigue strength according to Reference [1].

E. Niemi et al., *Structural Hot-Spot Stress Approach to Fatigue Analysis of Welded Components*, IIW Collection, DOI 10.1007/978-981-10-5568-3_6

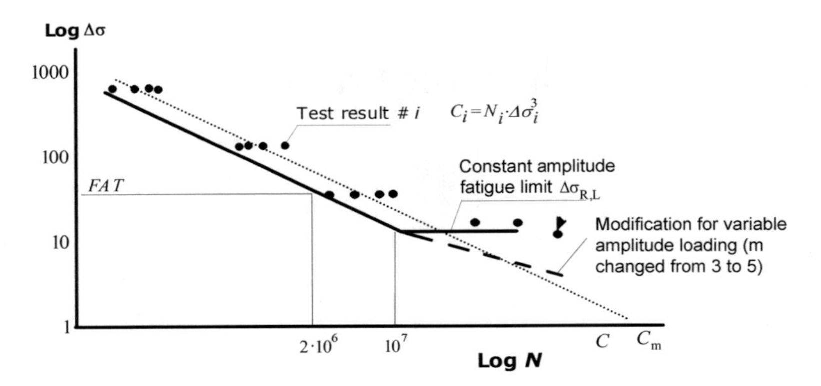

Fig. 6.1 Schematic presentation of derivation of the design S-*N* curve from a sample of test results

Referring to Fig. 6.1, in the case of constant amplitude loading a fatigue limit, $\Delta\sigma_{R,L}$, appears in the *S-N* curve. If all applied stress ranges remain below the fatigue limit, no fatigue damage is assumed to occur and an infinite fatigue life is expected. The constant amplitude fatigue limit is also valid for variable amplitude loading, but only if all stress ranges are below the constant amplitude fatigue limit.

It should be mentioned here that the constant amplitude fatigue limit is currently under discussion because fatigue failures have been observed below this limit after very high number of cycles (10^9 and more). Meanwhile, it has been proposed [1] to neglect the current sharp cut-off fatigue limit at $N = 10^7$ cycles and instead extend the constant amplitude S-*N* curve beyond $N = 10^7$ cycles to 10^9 cycles with a slope exponent of $m = 22$.

A further change is made to the constant amplitude S-*N* curve for considering variable amplitude loading. In particular, it is extrapolated beyond the 'knee point' at $N = 10^7$ cycles with a slope exponent of $m = 5$ for use in Miner's rule.

In the conventional structural hot-spot stress approach, based on surface stress extrapolation, through-thickness linearization or single-point stress at a distance of $0.5t$, the design curves presented are applicable for the range of section thicknesses up to a reference thickness t_{ref}. In view of the size effect discussed earlier, they must be lowered when considering thicker sections by multiplying allowable hot-spot stress range obtained from the design curve by the following factor:

$$f(t) = \left(\frac{t_{ref}}{t_{eff}}\right)^n \tag{6.2}$$

where the reference thickness t_{ref} is given in 6.2.1, n is specified for different weld types in Table 6.1 and the effective thickness is defined as follows:

$L/t > 2$: $t_{eff} = t$
$L/t \leq 2$: $t_{eff} = 0.5 \cdot L$, or
$t_{eff} = t_{ref}$, whichever is the larger.

t and L see the Fig. 6.2.

Table 6.1 Hot spot S-N curves for plates made of steel and aluminium alloys up to 25 mm thick

No	Joint	Description	Quality	FAT_{St}	FAT_{Al}	n
1		Butt joint	As-welded, proved free from significant flaws by NDT	100	40	0.2
2		Cruciform or T-joint with full penetration K-butt welds	K-butt welds, no lamellar tearing	100	40	0.3
3		Non-load carrying fillet welds	Trans. non-load carrying attachment, not thicker than the main plate, as-welded			
4		Bracket or stiffener ends, welds either welded around or not	Fillet weld(s) as-welded			
5		Cover plate ends and similar joints				
6		Cruciform joint with load-carrying fillet welds	Fillet weld(s) as-welded	90	36	0.3
7		Lap joint with load-carrying fillet welds				
8	L ≤ 100 mm	Type "b" joint with short edge attachment	Fillet or full penetration weld, as-welded	100	40	0.1
9	L > 100 mm	Type "b" joint with long edge attachment	Fillet or full penetration weld, as-welded	90	36	0.1

Note 1 These curves are valid for perfectly aligned joints. Misalignment effects of more than 5% must be considered in the applied stress range

Note 2 The nominally non- or partially load-carrying fillet welds shown under no. 3 and 5 may actually be load-carrying in certain cases, e.g. for very large attachments or if the bending of the base plate is restrained. In these cases load-carrying fillet welds should be assumed with FAT class 90 for steel and 36 for aluminium. This may also apply to no. 4 without soft bracket end

Note 3 A further reduction by one FAT class is recommended for fillet welds having throat thicknesses of less than one third of the thickness of the base plate

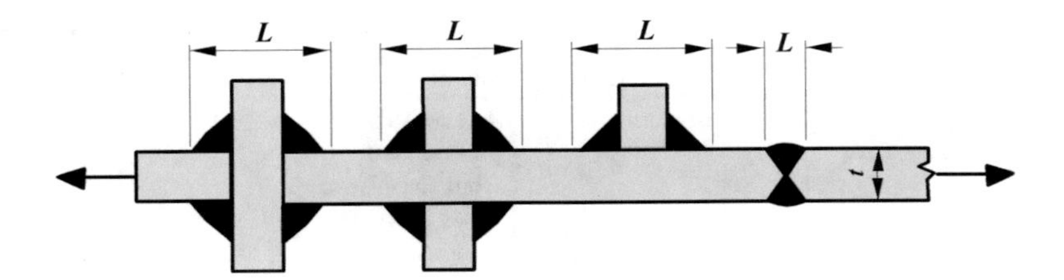

Fig. 6.2 Explanation of effective thickness

The plate thickness effect is already considered in the structural stress approaches by Dong [2], Xiao and Yamada [3] and Haibach [4] so that Eq. (6.2) is not applicable there.

In the case of load-carrying fillet welds, a separate fatigue analysis considering potential fatigue failure in the weld throat from the root should be performed. The nominal stress method, effective notch stress method or crack propagation analysis based on fracture mechanics are suitable approaches [1]. Also approaches based on structural stress in the weld throat have been developed [5].

6.2 Recommended *S-N* Curves for the Conventional Structural Hot-Spot Stress Approach

6.2.1 Hot-Spot S-N Curves

Based mainly on published data for steel welded specimens and structural components [6, 7], in which the data could be expressed in terms of the structural hot-spot stress range, the design *S-N* curves given in Table 6.1 are recommended for different weld types and quality. The difference between the two FAT classes proposed is mainly due to the more pronounced force flow around the toe of load carrying fillet welds. This might also occur in joints where the force is mainly transmitted by the weld to an attachment. In cases where the structural hot-spot stress is not based on surface stress extrapolation or through-thickness linearization, but on a single point at a distance of $0.5t$ from the hot spot, a reduction of the fatigue strength by one FAT class is recommended [7]. The data for welded aluminium alloys are based on references [1, 8–10]. In every case, the slope m in Eq. (3.1) is 3.0. Furthermore, the reference thickness t_{ref} in Eq. (6.2) is 25 mm which is also recommended for aluminium alloys [1]. Values of the exponent n to be used in Eq. (6.2) are included in Table 6.1.

It should be noted that the *S-N* curves are only valid for well aligned joints. Thus, any misalignment effects above small tolerances must be included in the calculated structural hot-spot stress range, see also 2.4.3. Also, the curves only apply for operating temperatures below 150 °C in steel and 50 °C in aluminium alloys.

6.2.2 Hot-Spot S-N Curves for Tubular Joints in Steel

Special hot-spot stress design *S-N* curves have been recommended by the IIW [11]. These are applicable to circular and rectangular hollow section joints. The thickness effect seen in tubular joints is rather pronounced and, moreover, the slope of the *S-N* curve *m* depends on the tube wall thickness. These effects may come from the fact that shell bending stresses dominate in the chord member. The tubular joint design curves should not be applied to other types of structure.

6.3 Recommended S-N Curves for the Other Structural Stress Approaches

6.3.1 Structural Stress Approach According to Dong

In the structural stress approach proposed by Dong [2], the design *S-N* curve is defined using a special structural stress parameter ΔS_s which is defined as follows:

$$\Delta S_s = \Delta\sigma_s \cdot t^{\frac{m-2}{2m}} / I(r)^{\frac{1}{m}} \quad (6.3)$$

where $\Delta\sigma_s$ structural stress range
t plate thickness
m exponent (m = 3.6)
r degree of bending (bending portion to total structural stress)
I(r) fatigue life integral from crack propagation analysis considering effects of degree of bending, crack front (semi-elliptical crack or 2D crack) and boundary condition (load- or displacement controlled)

The fatigue life integral varies between 1.1 and 1.3 for load controlled cases depending of the degree of bending and the shape of crack front, derived on the basis of two-stage crack growth model incorporating short anomalous regime. Further information can be found in [12, 13].

If the equivalent structural stress parameter is used in Eq. (6.1), the following constants define the design *S-N* curve derived from a large number of tests [12]:

$$C = (13{,}876\ \text{MPa})^{m'} \quad m' = 3.125 \quad (6.4)$$

Misalignment effects are already considered in the design *S-N* curve as far as these were present in the prevailing tests, but precise details are not known. Plate thickness effects are already contained by Eq. (6.3) as mentioned above.

6.3.2 Structural Stress Approach According to Xiao and Yamada

In the structural stress approach proposed by Xiao and Yamada [3], a design *S-N* curve is proposed which corresponds to FAT 100 defined in 6.1.

As mentioned before, the plate thickness effect is included in the structural stress obtained at a depth of 1 mm on the expected crack path. Unless included in the FE model, misalignment effects are not explicitly considered, which means that they may have to be taken into account as outlined in 2.4.3.

6.3.3 Structural Stress Approach According to Haibach

In the approach proposed by Haibach [4], which is based on the cyclic structural strain at a distance of about 2 mm from the weld toe, the mean endurable strain at two million cycles has been found to be between 0.06 and 0.14% for a stress ratio $R = -1$ [14]. Regarding plate thickness and misalignment effects, the same aspects as mentioned in 6.3.2 apply.

References

1. Hobbacher, A.F.: Recommendations for Fatigue design of welded joints and components, 2nd edn, Springer (2016)
2. Dong, P.: A Structural stress definition and numerical implementation for fatigue analysis of welded joints. Intl. J Fatigue **23**, 865–876 (2001)
3. Xiao, Z.G., Yamada, K.: A method of determining geometric stress for fatigue strength evaluation of steel welded joints. Int. J. Fatigue **26**, 1277–1293 (2004)
4. Haibach, E.: Die Schwingfestigkeit von Schweissverbindungen aus der Sicht einer örtlichen Beanspruchungsmessung (The fatigue strength of welded joints considered on the basis of a local stress measurement), LBF-Bericht FB-77. Fraunhofer-Inst. für Betriebsfestigkeit, Darmstadt (1986)
5. Fricke, W.: IIW guideline for the assessment of weld root fatigue. Welding in the World **57**, 753–791 (2013)
6. Partanen, T., Niemi, E.: Hot spot *S-N* curves based on fatigue tests of small MIG-welded aluminium specimens. IIS/IIW-1343-96 (ex. doc. XIII-1636-96/XV-921-96), Welding in the World, **43**(1), 16–23 (1999)
7. Maddox, S.J.: Hot-spot stress design curves for fatigue assessment of welded structures, Intl. J Offshore and Polar Engng., 12, No. 2, June 2002, pp 134-141
8. Partanen, T., Niemi, E.: Collection of hot-spot *S-N* curves based on tests of small arc-welded steel specimens, IIW Doc XIII-1602-99 (1999)
9. Maddox, S.J.: Hot-spot fatigue data for welded steel and aluminium as a basis for design, IIW Document No. XIII-1900a-01 (2001)
10. Tveiten, B.W.: Fatigue assessment of welded aluminium ship details. Doctoral thesis, Department of Marine Structures, Norwegian University of Science and Technology, Trondheim (1999)

11. Zhao, X.-L., Packer, J.A.: Fatigue design procedure for welded hollow section joints, Recommendations of IIW Sub-commission XV-E. Woodhead Publishing, Cambridge (2000)
12. Dong, P., et al.: The master S-N curve method: an implementation for fatigue evaluation of welded components in the 2007 ASME B&PV Code, Section VIII, Division 2 and API 579-1/ASME FFS-1, WRC Bulletin, No. 523, New York, 2010
13. Dong, P., Hong, J.K., Cao, Z.: Stresses and stress intensities at notches: 'anomalous crack growth' revisited. Int. J. of Fatigue **25**, 811–825 (2003)
14. Radaj, D., Sonsino, CM., Fricke, W.: Fatigue assessment of welded joints by local approaches, 2nd edn. Woodhead Publ., Cambridge (2006)

Chapter 7
Case Study 1: Box Beam of a Railway Wagon

7.1 Introduction

This example is based on a failure case study. Cracks were observed in the main girder of a railway wagon with a box section fabricated from structural steel. Strain gauges were attached to the structure and strain histories were recorded during a trip of 114 km in normal use. The goal of the study was to find the cause of the cracking. To this end, the strain measurements were interpreted in terms of the structural hot-spot stress.

7.2 Materials and Methods

7.2.1 *Description of the Structure*

Figure 7.1 shows the middle part of the wagon frame. The web of the box section was 4 mm thick. Cross-beams were fillet welded to the web. The bottom flange of the crossbeam was a flat bar 100 × 6 mm. There were diaphragm plates inside the box, as well as horizontal flat bar stiffeners corresponding to the bottom flange of the crossbeam. The hot-spot was located in the web at the toe of the fillet welds welded around the edge of the bottom flange and the flat bar stiffeners.

7.2.2 *Angular Misalignment in the Web*

The double fillet welds between the diaphragm and the web caused significant distortion of the web. The middle point of the web panel was displaced 5 mm inwards. Therefore, the hot-spot was located on the inner surface of the web.

E. Niemi et al., *Structural Hot-Spot Stress Approach to Fatigue Analysis of Welded Components*, IIW Collection, DOI 10.1007/978-981-10-5568-3_7

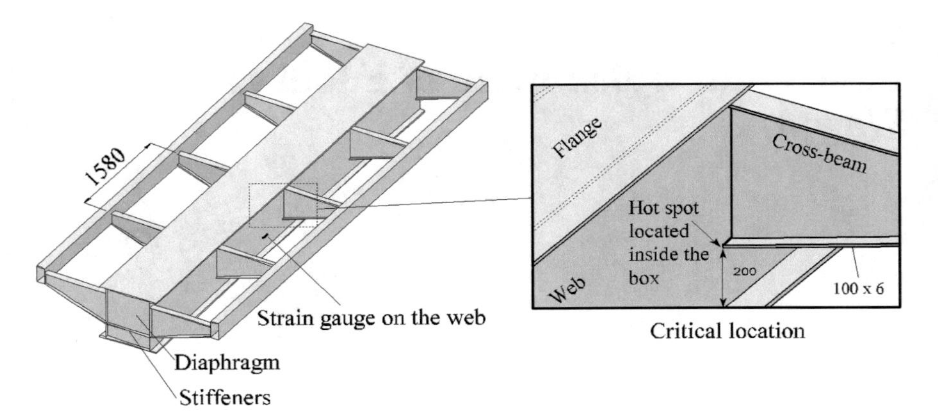

Fig. 7.1 Section of the wagon frame showing the hot-spot location

7.2.3 Strain Gauge Measurements

The structural hot-spot stress could not be measured directly because it was located on the inside of the box. One strain gauge was attached on the top of the box girder and one on the bottom, in the web plane. Moreover, one gauge was placed on the outer surface of the web between two cross-beams at the level of their bottom flanges, Fig. 7.1.

Comparison of the strains from the above-mentioned gauges showed that the gauge in the web gave a good indication of the nominal bending stress in the girder in the vicinity of the hot-spot. Rainflow cycle-counting analysis of the stress history corresponding to the 114 km trip gave the nominal stress range occurrences at the critical locations, expressed in terms of 32 levels, at the critical location shown in Table 7.1.

Table 7.1 Measured nominal stress range occurrences corresponding to a trip of 114 km. The trip included one loading/unloading cycle

i	$\Delta\sigma_i$ (MPa)	n_i (cycles)	i	$\Delta\sigma_i$ (MPa)	n_i (cycles)	i	$\Delta\sigma_i$ (MPa)	n_i (cycles)	i	$\Delta\sigma_i$ (MPa)	n_i (cycles)
1	73.5	0	9	55.1	4	17	36.7	6	25	18.3	353
2	71.2	0	10	52.8	3	18	34.4	11	26	16.0	270
3	68.9	0	11	50.5	4	19	32.1	19	27	13.7	333
4	66.6	1	12	48.2	8	20	29.8	43	28	11.4	813
5	64.3	0	13	45.9	10	21	27.5	116	29	9.1	1055
6	62.0	5	14	43.6	5	22	25.2	274	30	6.8	2924
7	59.7	0	15	41.3	9	23	22.9	302	31	4.5	6440
8	57.4	2	16	39.0	8	24	20.6	332	32	2.2	76578

7.2.4 Structural Hot-Spot Stress Determination

The region of interest was the junction of the bottom flange of the cross-beam and the stiffeners inside the box. Because a suitable parametric formula for the resulting membrane stress concentration factor could not be found, a local solid element model, subjected to constant membrane stress, was used to calculate it. It was thought that the actual geometry could be modelled as the simpler case of a loaded plate with longitudinal gussets welded to each surface, as shown in Fig. 7.2. This constant stress case was deemed sufficient although the actual membrane stress distribution in the web follows from beam bending. It was only necessary to model 1/8 of the detail, due to threefold symmetry. A region with the following dimensions was analysed: $L = 800$ mm, $l_1 = 100$ mm, $h_1 = 100$ mm, $t = 4$ mm, $t_1 = 6$ mm, $b = 400$ mm. Linear solid elements were used for proper modelling of the weld fillet. Figure 7.3 shows a part of the model in the vicinity of the hot-spot. Due to the three symmetry planes, only one half of the web plate, 2 mm thick, was modelled. The model was loaded at the end to give a constant membrane stress of 1 MPa. The maximum membrane stress at the centre-line of the detail was determined. This was obtained as the average of the symmetry plane and surface stresses at distances of 1.6 and 4.0 mm from the weld toe. Linear extrapolation to the weld toe was then performed according to Eq. (3.1) to determine the structural hot-spot stress.

The stress magnification due to angular misalignment caused by welding distortion was calculated using Eqs. (5.5) and (5.6). The length term $2\,l$ was assumed to correspond to the spacing of the cross-beams, 1580 mm, and 'fixed ends' boundary condition was assumed due to symmetry.

The stress magnification due to angular misalignment is non-linear with respect to applied loading, since the misalignment decreases as the joint is loaded. Therefore, K_m was estimated for both the maximum and minimum stresses at the most damaging stress level separately, including allowance for the second-order beneficial effect of straightening using Eq. (5.6).

A question arises as to whether the stress magnification factor should be applied to the nominal membrane stress, $\sigma_{m,nom}$, or to the more local membrane stress

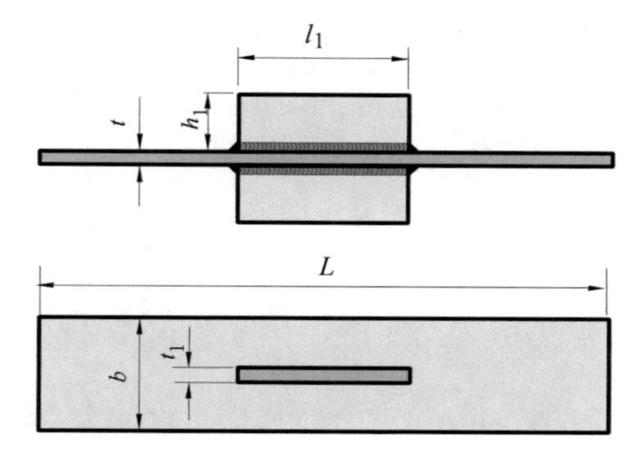

Fig. 7.2 A detail with double gussets for calculation of the membrane stress concentration

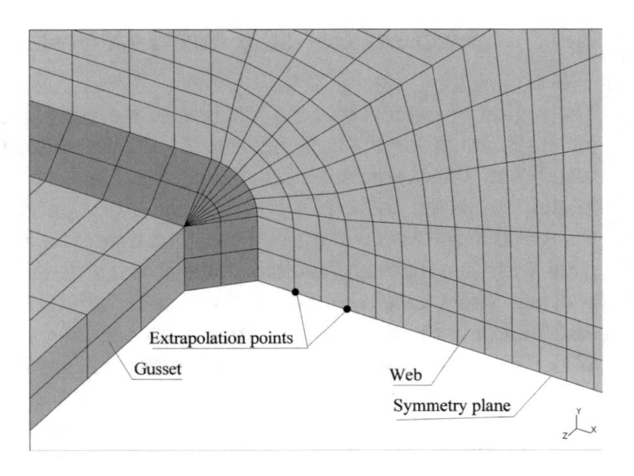

Fig. 7.3 Hot-spot region of the FE model. A relatively fine element mesh was planned due to the use of linear 8-node solid elements. Due to the plane of symmetry, the total thickness of the web consisted of two element layers

increased by the stress concentration effect of the structural detail. In this particular case, the stress concentration occurs in a rather narrow area, and the secondary shell bending induced by straightening of the web plate could be considered as a more global phenomenon. Therefore, it is sufficient to calculate the secondary bending stress $(K_m - 1)\,\Delta\sigma_{m,nom}$ separately, and add it to the increased membrane stress. Thus, the structural hot spot stress in this case is given by:

$$\Delta\sigma_s = (K_s + K_m - 1) \cdot \Delta\sigma_{m,nom} \tag{7.1}$$

For simplicity, the magnification factor was resolved only for the most damaging stress range, and the same factor was then applied to all stress ranges. Therefore, the damage distribution was first calculated assuming no misalignment effect.

7.2.5 S-N Curve

The structural hot-spot stress *S-N* curve for non-load-carrying fillet welds, FAT 100, from Table 6.1, was used. In order to avoid excessive conservatism, the Miner's damage sum was based on a bilinear *S-N* curve assuming a slope exponent change from $m = 3$ to 5 at the knee point corresponding to $N = 10^7$ cycles. Thus, the following values for the *S-N* curve according to Eq. (6.1) are applied, where index "1" refers to the curve above and index "2" beyond the knee point:

$$C_1 = 2 \cdot 10^{12}$$
$$C_2 = 6.8514 \cdot 10^{15}$$
$$m_1 = 3$$
$$m_2 = 5$$

The stress range at the knee point is $\Delta\sigma_{th} = 58.5$ MPa.

7.2.6 *Partial Safety Factors*

In design calculations, appropriate partial safety factors should be used. Then the stress ranges are multiplied by γ_F and the fatigue strength values are divided by γ_M. However, the goal of the present study was to predict the characteristic life of the detail. Therefore

$$\gamma_F = 1.0$$
$$\gamma_M = 1.0.$$

7.3 Results

7.3.1 *Stress Concentration Factor, K_s*

The finite element model yielded the following results:

$$\sigma_m(1.6\,\text{mm}) = 1.442\,\text{MPa}$$
$$\sigma_m(4.0\,\text{mm}) = 1.283\,\text{MPa}$$

Applying Eq. (3.1) gives:

$$\sigma_{hs} = 1.67 \cdot 1.442 - 0.67 \cdot 1.283 = 1.548\,\text{MPa}$$

or, rounding up conservatively:

$$K_s = 1.6$$

7.3.2 *Results for a Perfectly Straight Web*

Table 7.2 shows the damage calculation when no misalignment is taken into account. The nominal stress ranges have been multiplied by $K_s = 1.6$, giving structural hot-spot stress ranges $\Delta\sigma_{i,s}$. The three lowest stress levels have been omitted in this case.

The total damage is $D = 32 \times 10^{-6}$. The predicted life in this case becomes:

$$N = \frac{114\text{ km}}{32 \cdot 10^{-6}} = 3{,}562{,}500\text{ km}$$

Table 7.2 Damage calculation assuming no misalignment. The bold line represents the knee point of the *S-N* curve

Level i	$\Delta\sigma_{i,nom}$	$\Delta\sigma_{i,s}$	n_i	$N_{fi} \times 10^{-6}$	$D_i \times 10^6$
4	66.6	106.6	1	1.65	0.6050
6	62.0	99.2	5	2.05	2.4405
8	57.4	91.8	2	2.58	0.7746
9	55.1	88.2	4	2.92	1.3704
10	52.8	84.5	3	3.32	0.9044
11	50.5	80.8	4	3.79	1.0550
12	48.2	77.1	8	4.36	1.8347
13	45.9	73.4	10	5.05	1.9805
14	43.6	69.8	5	5.89	0.8487
15	41.3	66.1	9	6.93	1.2984
16	39.0	62.4	8	8.23	0.9719
17	36.7	58.7	6	9.88	0.6074
18	34.4	55.0	11	13.61	0.8080
19	32.1	51.4	19	19.10	0.9949
20	29.8	47.7	43	27.75	1.5495
21	27.5	44.0	116	41.54	2.7922
22	25.2	40.3	274	64.45	4.2511
23	22.9	36.6	302	104.32	2.8949
24	20.6	33.0	332	175.07	1.8964
25	18.3	29.3	353	317.28	1.1126
26	16.0	25.6	270	623.13	0.4333
27	13.7	21.9	333	1360.06	0.2448
28	11.4	18.2	813	3431.01	0.2370
29	9.1	14.6	1055	10328.0	0.1021
					32.0083

7.3.3 *Effective Magnification Factor,* K_m

Table 7.2 shows that most damage occurred at stress level number 22 with $\Delta\sigma_{nom}$ = 25.2 MPa. Assuming that the mean nominal stress is 40 MPa, the maximum and minimum stresses for this stress range are:

$$\sigma_{nom,max} = 52.6\,\text{MPa}$$
$$\sigma_{nom,min} = 27.4\,\text{MPa}$$

For the maximum stress, Eqs. (5.5 and 5.6) yield:

$$\beta = \frac{1580}{4} \cdot \sqrt{\frac{3 \cdot 52.6}{210000}} = 10.83$$

$$K_{\mathrm{m}} = 1 + \frac{3 \cdot 5}{4} \cdot \frac{\tanh 5.415}{5.415} = 1.692$$

$$\sigma_{\max} = 1.692 \cdot 52.6\mathrm{MPa} = 89.0\mathrm{MPa}$$

and for minimum stress

$$\beta = \frac{1580}{4} \cdot \sqrt{\frac{3 \cdot 27.4}{210000}} = 7.815$$

$$K_{\mathrm{m}} = 1 + \frac{3 \cdot 5}{4} \cdot \frac{\tanh 3.908}{3.908} = 1.959$$

$$\sigma_{\min} = 1.959 \cdot 27.4\mathrm{MPa} = 53.7\mathrm{MPa}$$

The magnified stress range is then 89.0 − 53.7 = 35.3 MPa
This leads to an effective magnification factor:

$$K_{\mathrm{m,e}} = \frac{35.3}{25.2} = 1.402$$

7.3.4 Results for a Web with Angular Misalignment

Table 7.3 shows the results when all stress ranges are corrected according to Eq. (7.1).

The total damage during the test trip now becomes $D = 80.53 \times 10^{-6}$. The predicted characteristic life for the misaligned web becomes:

$$N = \frac{114\ \mathrm{km}}{80.53 \cdot 10^{-6}} = 1{,}415{,}600\ \mathrm{km}$$

7.4 Discussion and Conclusions

The predicted life, 1,415,600 km, corresponds rather well to the time when cracks were first observed. The effect of welding distortion in the web is fairly large. The calculations show that it reduces the life by 60%.

The second order factor according to Eq. (5.5), $\tanh(\beta/2)/(\beta/2)$, in the distorted web had a significant effect. In such a case, it is not sufficient to apply the magnification factor, K_{m}, to the stress range directly (Eq. 2.2). Instead, it should be applied separately to maximum and minimum stresses in order to find out the actual effect on the stress range. An effective magnification factor was determined only for

Table 7.3 Damage calculation taking the misalignment into account. The bold line represents the knee point of the *S-N* curve

Level i	$\Delta\sigma_{i,nom}$	$\Delta\sigma_{i,s}$	n_i	N_{fi} x 10^{-6}	D_i x 10^6
4	66.6	133.3	1	0.84	1.1852
6	62.0	124.1	5	1.05	4.7809
8	57.4	114.9	2	1.32	1.5175
9	55.1	110.3	4	1.49	2.6846
10	52.8	105.7	3	1.69	1.7717
11	50.5	101.1	4	1.94	2.0668
12	48.2	96.5	8	2.23	3.5941
13	45.9	91.9	10	2.58	3.8797
14	43.6	87.3	5	3.01	1.6626
15	41.3	82.7	9	3.54	2.5436
16	39.0	78.1	8	4.20	1.9039
17	36.7	73.5	6	5.04	1.1899
18	34.4	68.9	11	6.12	1.7965
19	32.1	64.3	19	7.54	2.5213
20	29.8	59.7	43	9.42	4.5654
21	27.5	55.1	116	13.49	8.5988
22	25.2	50.5	274	20.86	13.1349
23	22.9	45.8	302	34.00	8.8829
24	20.6	41.2	332	57.72	5.7523
25	18.3	36.6	353	104.32	3.3838
26	16.0	32.0	270	204.19	1.3223
27	13,7	27,4	333	443.64	0.7506
28	11,4	22,8	813	1112.0	0.7311
29	9,1	18,2	1055	3431.0	0.3075
					80.5279

the most damaging stress range, and it was applied to all stress range levels for simplicity.

The fact that the misalignment effect raised the stress range at three strongly damaging levels, 20, 21 and 22, closer to the knee point of the *S-N* curve, further increased the calculated damage.

Chapter 8
Case Study 2: Hatch Corner Design for Container Ships

8.1 Introduction

This example comes from the investigation of the fatigue strength capacity of an advanced, production-friendly hatch corner design for container ships. The proposed change introduced a new hot-spot, which may reduce the fatigue capacity of the hatch corner. The goal of the study was to verify the applicability of the structural hot-spot method to this detail by component testing.

8.2 Materials and Methods

8.2.1 Description of the Structure

Figure 8.1 shows both the conventional and proposed new advanced designs. In the conventional design, the deck plating and the transverse deck strip between adjacent holds consisted of a continuous plate with a transition radius at the hatch corner. In the advanced, production-friendly design, longitudinal bulkhead and hatch coaming form a continuous plate to which the separate transverse deck strip is welded. In this way, fabrication is facilitated, especially in the final stages of assembly. Unfortunately, a new hot-spot is formed at the end of the transverse deck strip.

8.2.2 Service Loads

The structure in way of the hatch corner is subjected to axial stresses in the deck plating due to longitudinal hull bending. Further, warping of the hull sections caused by hull torsion creates bending moments around the vertical axis in the

E. Niemi et al., *Structural Hot-Spot Stress Approach to Fatigue Analysis of Welded Components*, IIW Collection, DOI 10.1007/978-981-10-5568-3_8

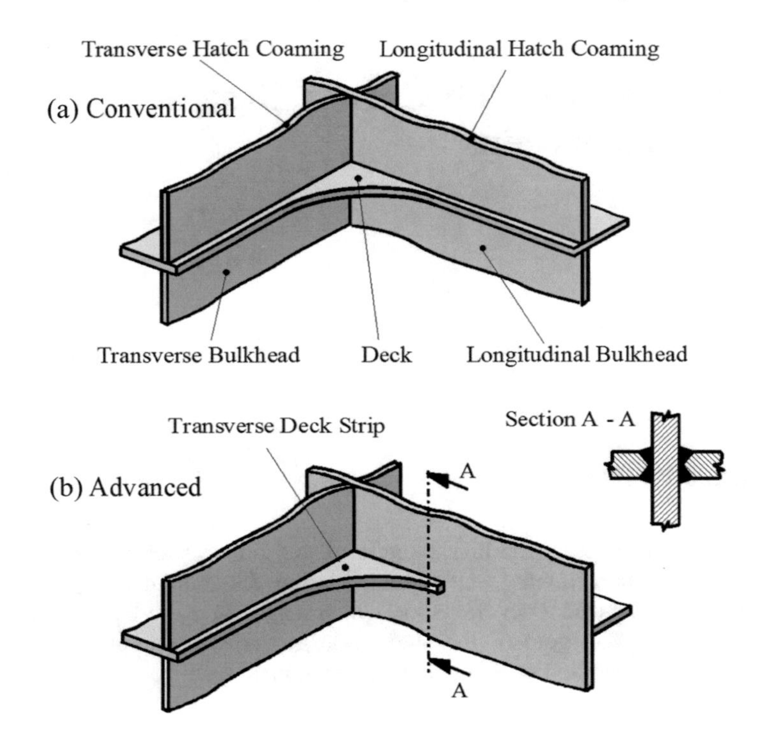

Fig. 8.1 Hatch corner details. The hot spot studied is located close to Section A-A

transverse deck strip. Both types of load consist of a still water and a wave-induced component. The total number of wave-induced stress cycles is about 50 million in a 20 year service life, with a log-linear distribution of stress ranges.

8.2.3 Experimental Investigation

The service load condition was modelled in a three-point bending test as shown in Fig. 8.2. The test specimen represented the advanced hatch corner design. In the transverse direction, only the deck strip was modelled, while in the longitudinal direction the girder web represented the deck plate and the girder flange adjacent to the transverse deck strip represented the longitudinal bulkhead/hatch coaming. Initially, the fatigue-critical zone was designed as shown in "Design A". It contained both type "a" and type "b" hot spots. However, in the test pieces fabricated by a shipyard the weld produced was as shown in "Design B". Consequently, only the type "a" hot-spot remained.

Prior to fatigue testing, strain gauge measurements were carried out under static load. Fatigue tests were then performed under constant amplitude load at two

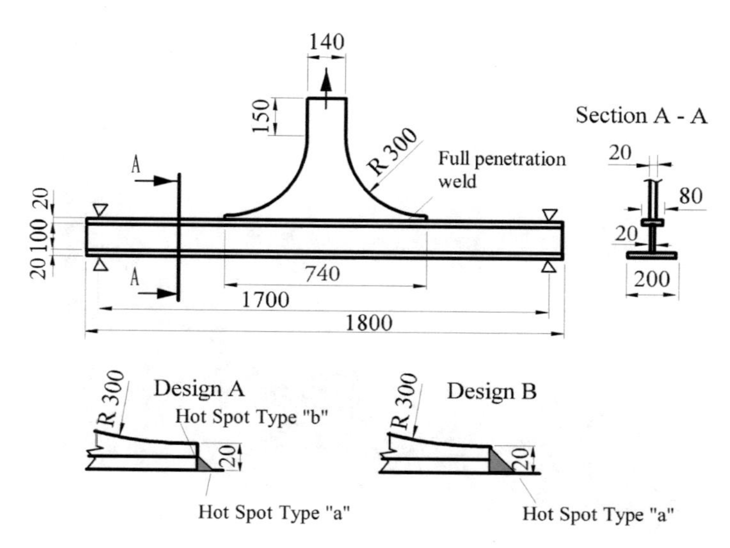

Fig. 8.2 Three-point-bending test specimen

load levels, at a stress ratio $R = 0$ (zero to tension load) and at a test frequency of about 25 Hz.

8.2.4 Structural Hot-Spot Stress Determination

The stress distribution in the test specimen was analysed by means of a finite element model (Fig. 8.3). Solid 20-node elements with reduced integration order were applied and use was made of the symmetry around the half-length of the specimen. The loading conditions considered included axial load (LC1) and three-point bending (LC2). Strain gauge measurements were made for verification of the stress distribution under LC2. Due to the slenderness of the upper flange, only uniaxial strains were measured. Stresses derived from the strain measurements are plotted in Fig. 8.4, together with calculated stresses. The differences between measured and calculated stresses are typical for the element mesh used, resulting from the singular behaviour at the weld toe. Due to the use of the recommended extrapolation points the extrapolation method yielded satisfactory estimates of the actual structural hot-spot stress.

Structural hot-spot stresses were obtained by means of linear extrapolation according to Eq. (4.1). The stresses shown in Fig. 8.4 are non-dimensional, related to the nominal axial or bending stress in the cross-section at the hot spot, as appropriate. Structural stress concentration factors of $K_s = 1.52$ for axial load (LC1) and $K_s = 1.60$ for three point bending (LC2) were obtained.

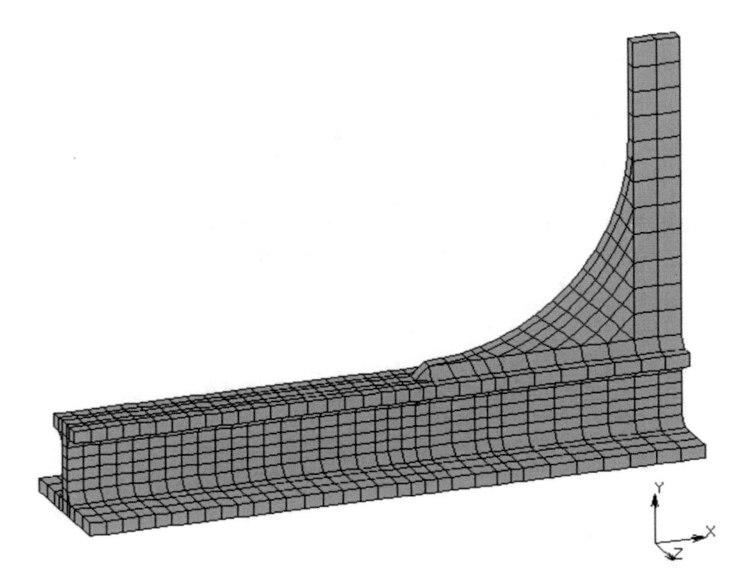

Fig. 8.3 Solid element model used in structural hot spot stress analysis, according to Fig. 4.3b

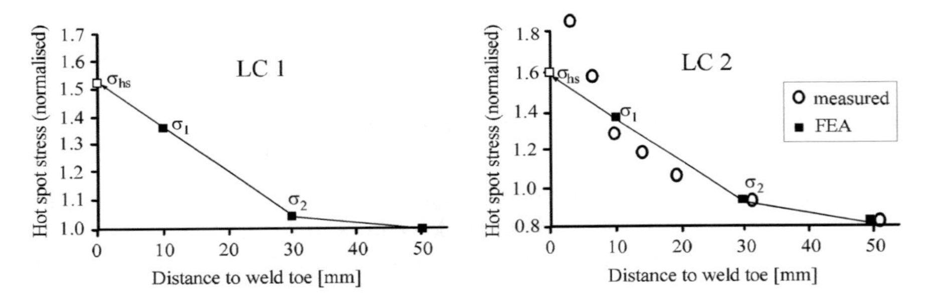

Fig. 8.4 Results of finite element analyses for axial load (LC1) and three-point bending (LC2)

8.2.5 S-N Curve Based on Nominal Stress

The results of the fatigue tests are shown in Fig. 8.5, based on the nominal stress range as calculated by means of simple beam theory. The fatigue life of each test piece was taken as the number of cycles until shut-off of the testing machine, which took place when the testing frequency fell by 0.05 Hz as a result of the increased displacement of the cracked specimen. The corresponding crack length on the surface of the beam flange (Fig. 8.1) varied between 15 and 60 mm. From the statistical analysis, an *S-N* curve with a slope of $m = 3$ equivalent to FAT 63 was derived, that is the nominal stress range was $\Delta\sigma_{nom} = 63$ MPa at 2 million cycles, for a probability of survival of 97.5%.

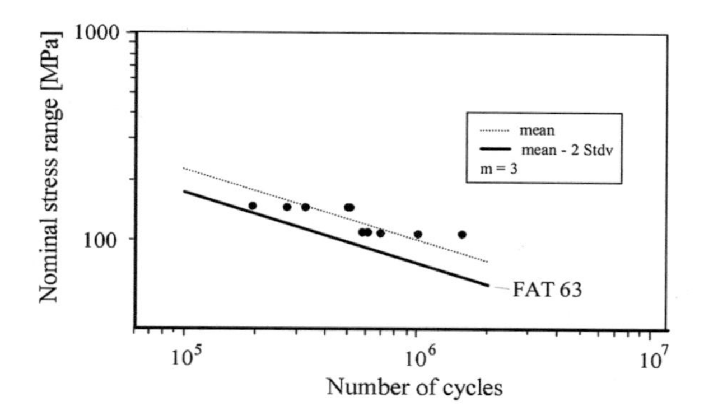

Fig. 8.5 Test results presented as nominal stress range versus fatigue life. Mean minus two standard deviations corresponds to fatigue class FAT 63

8.3 Fatigue Assessment

The results of the investigation were assessed according to Germanischer Lloyd rules [1]. For LC2, this gives a design curve with a reference structural hot-spot stress range of 100 MPa (*FAT* 100).

Assuming that the test results represent the fatigue behaviour of the full-scale structure, for LC2 loading the present investigation yields a fatigue class of:

$$\begin{aligned}\Delta\sigma_{\mathrm{hs}} &= K_{\mathrm{s}} \cdot \Delta\sigma_{\mathrm{nom}} \\ &= 1.60 \cdot 63 = 101\,\mathrm{MPa}\end{aligned}$$

In other words, the tests suggest a structural hot-spot stress design class of *FAT* 101, which is slightly higher than, and therefore consistent with the Germanischer Lloyd class of *FAT* 100.

8.4 Conclusion

A fatigue investigation based on hot-spot stresses was carried out for a proposed advanced design of hatch corners in ships. The results were in good agreement with the assessment method given in the rules of Germanischer Lloyd [1].

Reference

1. G.L.: Rules for classification and construction, Part I, Ship technology, 1.1 Seagoing ships—1 Hull, Part V, Analysis techniques, edn 1998. Germanischer Lloyd, Hamburg (1998)

Chapter 9
Case Study 3: Web Frame Corner

9.1 Introduction

Web frames in ships and comparable components in other structures are frequently designed as deep-web girders. At corners and intersections, the girders cross each other and the flanges form cruciform joints usually requiring full-penetration welds.

Such a situation occurs in 'roll-on/roll-off' ships, where the corners connecting the web frames of the ship's side and deck may be subjected to high cyclic stresses. Therefore, fatigue tests have been performed on three full-scale models [1], and these are selected for this case study.

The test model is shown in Fig. 9.1. It contains two 600 mm deep web frames. A diagonally acting hydraulic cylinder (Fig. 9.1b) creates a bending moment together with axial and shear forces in both frames. The forces produce relatively small stresses so that the frame is mainly subjected to bending.

In total three models were built, two with the usual configuration, while the third contained additional diagonal stiffeners on the web in the intersection area which were connected with full penetration welds at their ends. The material was higher-tensile ship structural steel A36, having a nominal yield stress R_{eH} = 355 N/mm^2.

All models were fabricated in the usual way by first placing the individual frames on the plates and then connecting the 'block joint' (between the white and coloured parts in Fig. 9.1a). Here, the 10 mm thick web was MAG-welded first and then the flange 200 × 20 was connected by full-penetration welding. This cruciform joint was expected to be the fatigue-critical connection.

This is a typical example of a transverse weld with structural stress concentration, similar to the transverse weld shown in Fig. 4.4a. Compared to the nominal stress in the flange, which is the sum of the axial and bending stresses in the web frame, an increased structural stress is expected in the centre of the weld due to the transfer of the bending moment from the horizontal into the vertical web frame.

E. Niemi et al., *Structural Hot-Spot Stress Approach to Fatigue Analysis of Welded Components*, IIW Collection, DOI 10.1007/978-981-10-5568-3_9

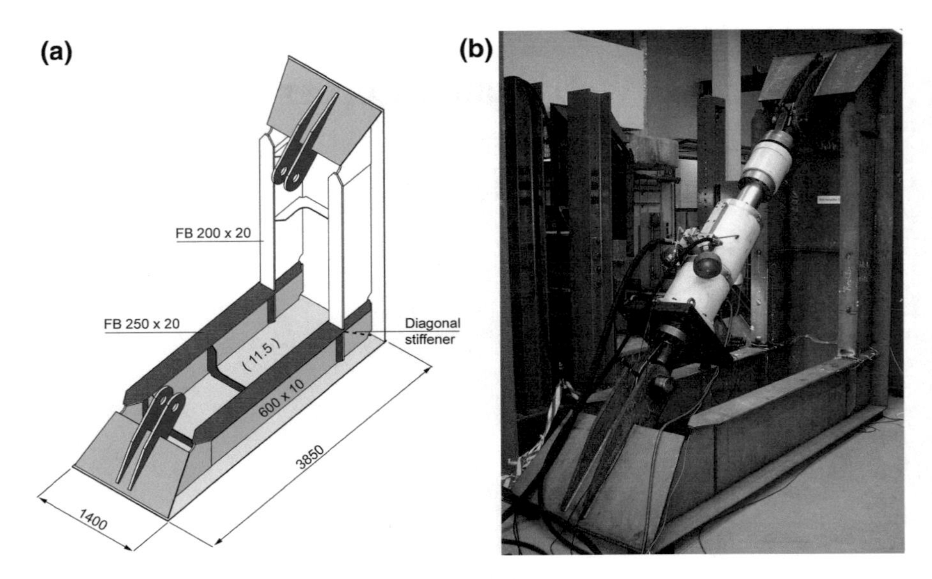

Fig. 9.1 Web frame corner investigated and test set-up

This results in a reduced effective breadth of the flange and, thus, a structural stress increase which can be calculated with the finite element method. The situation is further complicated by the inclusion of cope holes for welding.

9.2 Computation of the Structural Hot-Spot Stress

9.2.1 Finite Element Modelling

A finite element model of the test model was created, utilizing the vertical symmetry plane so that only half of the structure was modelled. Figure 9.2 shows the model together with the details of the critical connection, including the circular cope hole for welding the cruciform joint.

A fine mesh was chosen with further refinement in the longitudinal and vertical directions in comparison to Fig. 4.3a. 20-node solid elements with reduced integration were applied.

Figure 9.2 shows the refined mesh for the hot spot on the horizontal flange (HS1). Preliminary investigation identified three other critical hot-spots, HS2–HS4, which are included in Fig. 9.3. This also shows the nominal weld dimensions; these were used in the finite element model.

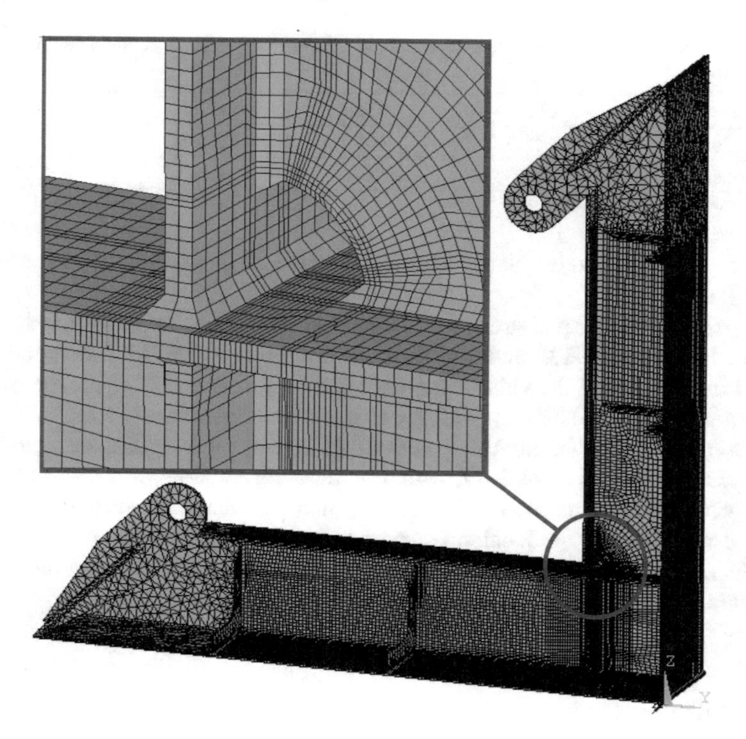

Fig. 9.2 Finite element model of the web frame corner

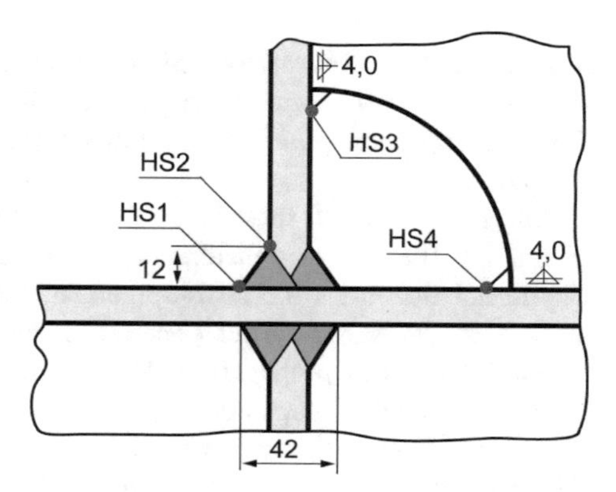

Fig. 9.3 Critical part of the web frame corner with hot-spots HS1–HS4

9.2.2 Computation of Structural Hot-Spot Stresses

The finite element model was subjected to a diagonal force of 100 kN at the hinges, corresponding to F = 200 kN for the test models incorporating two web frames. The nominal stress in the flange of the vertical web frame was $\sigma_{\text{nom}} = 71$ N/mm^2

Table 9.1 Computed structural hot-spot stress [N/mm^2] for the test models subjected to F = 200 kN

Hot-spot	HS1	HS2	HS3	HS4
Test model 1/2	145.5	244.0	203.5	90.4
Test model 3	128.6	207.6	189.3	68.2

assuming a fully effective flange and an effective plate width of 518 mm on the outer side.

The structural hot-spot stress was computed for all hot spots by linear extrapolation of the stresses $1.0t$ and $0.4t$ from the weld toe. Table 9.1 shows the results for test models 1 and 2 (without the diagonal stiffener) and 3 (with the diagonal stiffener, see Fig. 9.1a). Hot-spot HS2 was the most critical in all cases, followed by HS3. Compared with the nominal stress of 71 N/mm^2, the highest structural stress concentration factor K_s was 3.49, which is rather large. However, also notable are the reduced stresses in model 3, showing that the diagonal stiffener was quite effective for reducing the local stresses and avoiding plate buckling.

Axial misalignment of the vertical flanges was found to have negligible effect on the structural stresses so it was neglected.

9.3 Fatigue Tests

9.3.1 Performance of the Tests

The fatigue tests were performed under constant amplitude loading with a load ratio $R \approx 0$. The load range in the models was varied between ΔF = 230 kN and 280 kN. The corresponding nominal stress in the vertical flange ranged from 81.9 to 99.4 N/mm^2.

Surprisingly, in all models the first crack was not observed at hot-spot HS2, but at HS3. It then spent considerably time longer penetrating the flange thickness. Figure 9.4 illustrates the development of the crack length measured on the inner surface of the flange. After penetrating the flange (at a crack length of approx. 50 mm), the crack propagation rate rapidly increased.

A second crack appeared later at HS2 but only in the third model with the diagonal stiffener. It grew very quickly, leading to separation of more than half the flange width, such that it marked the end of the test.

9.3.2 Observed Fatigue Lives

For a comparison of the endured load cycles with design *S-N* curves, a specific failure criterion needs to be defined. This is usually through-section cracking,

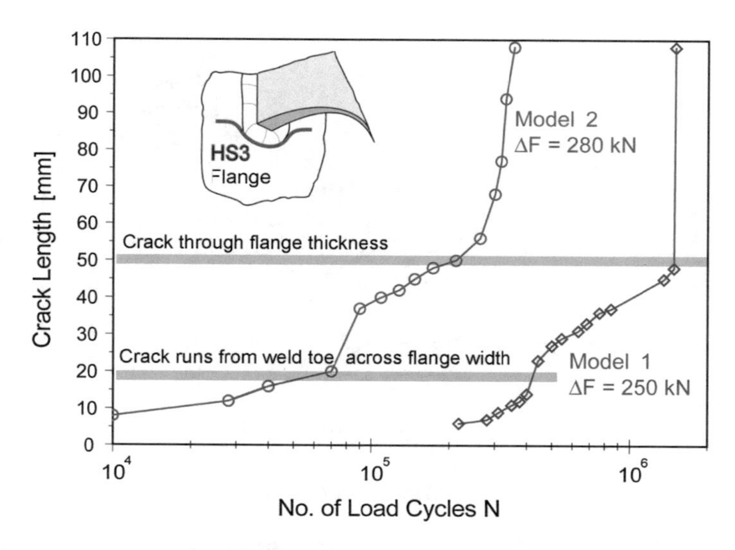

Fig. 9.4 Typical crack development at HS3

because a leak is present and the crack propagation rate is then considerably increased. The observed cracks penetrated the flange thickness when the surface length, as plotted in Fig. 9.4, was approximately 50 mm.

9.3.3 Comparison with Design S-N Curves

Using the computed nominal stress at the flange intersection and the structural hot-spot stress at HS3, the fatigue endurances from the tests are plotted in the *S-N* diagram in Fig. 9.5 for different crack lengths and compared with the relevant design *S-N* curves, FAT 71 for use with the nominal stress (cruciform joint) and FAT 90 for use with the structural hot-spot stress (load-carrying weld).

Considered in terms of the structural hot-spot stress, the results are on the safe side for the failure criterion chosen (crack length $\approx$ 50 mm). Hence, the tests confirm the approach. However, the results are mostly non-conservative when considered in terms of the nominal stress. The main reason for this is the considerably reduced effective breadth which increases the stress in the middle of the flanges as compared with the assumed nominal stress.

Finally, the reason why cracking first occurred at HS3 rather than HS2, where the structural hot-spot stress was higher, was investigated. It was found that the toe of the weld at HS2 had a larger radius than that at HS3, presumably due to gravity forces on the molten weld metal during welding. The corresponding lower local stress concentration is a possible explanation for the increased fatigue strength of the weld toe at HS2.

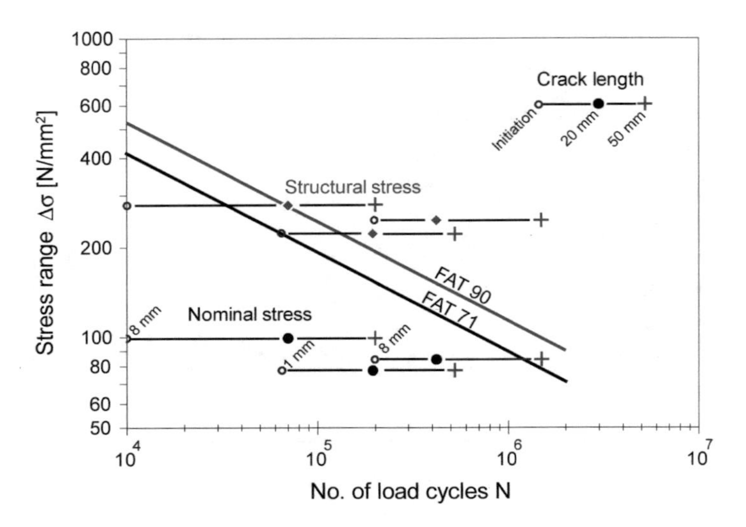

Fig. 9.5 S-*N* diagram for the critical web frame corners on the basis of nominal and structural hot-spot stresses, showing fatigue life for different crack lengths

Reference

1. Fricke, W., von Lilienfeld-Toal, A., Paetzold, H.: Fatigue strength investigations of welded details of stiffened plate structures in steel ships. Int. J. Fatigue **34**(1), 17–26 (2012)

Chapter 10
Case Study 4: Loaded Stiffener on T-Bar

10.1 Introduction

The arrangement of the stiffener on a T-bar is illustrated in Fig. 10.1a. Fatigue tests were performed by Kim et al. [1] using structural steel specimens with the dimensions shown in Fig. 10.1b. The stiffener was made from a flat bar and connected to the T-bar with 5 mm leg length fillet welds. The critical hot-spot, of Type "b", is encircled (detail "A").

The fatigue performance of the test specimen was evaluated on the basis of both the nominal and structural hot-spot stress. The structural hot-spot stress was computed for a nominal stress on the top edge of the stiffener at the weld toe of σ_{nom} = 100 N/mm^2 using surface stress extrapolation, Dong's method, assuming a final crack length of 10 mm, and Xiao/Yamada's method. The nominal stress-based fatigue evaluation was related to a nominal stress range $\Delta\sigma_{nom}$ = 100 N/mm^2. The present description of this rather simple example includes several details, including stresses at nodal points, that enable the reader to perform his own analysis and reproduce the results.

10.2 Computation of the Structural Hot-Spot Stress

10.2.1 Finite Element Modelling

A 3D shell model of the stiffener and the I-beam was created. Figure 10.2a shows the overall model, with shell elements arranged in the mid-thickness of the components. Shell elements with mid-side nodes were used, having a quadratic shape function. Figure 10.2b shows the modelling of detail "A". The weld was modelled

E. Niemi et al., *Structural Hot-Spot Stress Approach to Fatigue Analysis of Welded Components*, IIW Collection, DOI 10.1007/978-981-10-5568-3_10

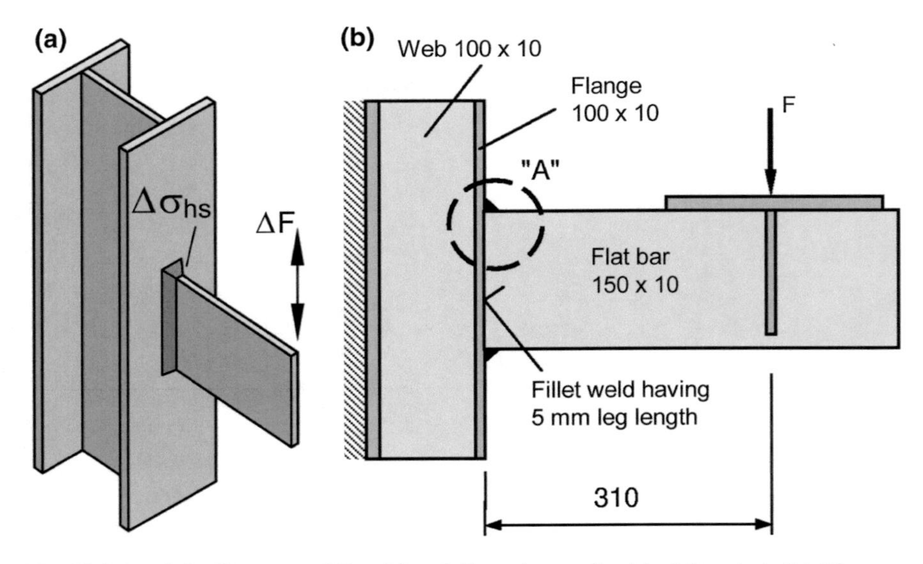

Fig. 10.1 Loaded stiffener on a T-bar (**a**) and dimensions realized in fatigue tests (**b**) [1]

in accordance with Fig. 4.4d, by adopting an increased thickness of 15 mm. Three meshes with varying element length of l = 10 mm, 5 mm and 2 mm were set-up. The critical area was further refined using 0.5 mm long elements for the application of Xiao/Yamada's method.

A force of $F = 12{,}100$ N was necessary to create a nominal stress $\sigma_{\mathrm{nom}} = 100$ N/mm^2 on the top edge of the stiffener at the intersection line between stiffener and I-beam.

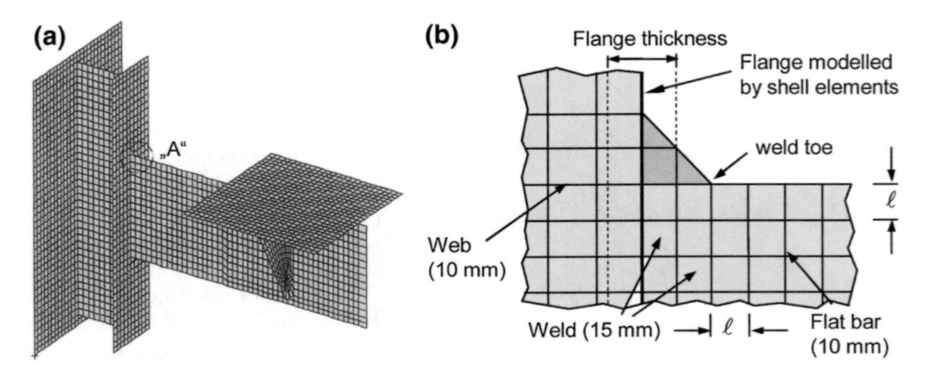

Fig. 10.2 **a** FE model of the loaded stiffener; **b** Elements at detail "A"

10.2.2 Determination of the Structural Hot-Spot Stress by Extrapolation

The structural hot-spot stress was first determined using the results of 10 mm long elements. This length matches the requirements stated in Fig. 4.3d for Type “b” hot spots. Figure 10.3 shows the longitudinal nodal stresses computed for the upper edges of the two elements next to the weld toe. These were obtained directly from the finite element program. The stresses at the mid-side nodes result from linear extrapolation of the stresses at the integration points (2 × 2 integration) to the element edge. As seen in Fig. 10.4, extrapolation using Eq. (4.2) yielded $\sigma_{hs} = 184.65$ N/mm^2, or $\sigma_{hs}/\sigma_{nom} = 1.8465$.

Figure 10.4 also shows the extrapolation (black arrow) together with the stress results obtained for the other element lengths used. It is interesting to note that, compared with the 10 mm element results, the fine mesh with $l = 2$ mm yields a distribution with lower stresses at a distance of about 5 mm from the weld toe. However, the quadratic extrapolation method (red arrow), proposed for fine meshes using reference points 4, 8 and 12 mm away from the weld toe (see Fig. 4.3b) yields almost the same result. As mentioned before, the reason is the slight stress overestimation in coarse meshes in the element next to the singularity of the weld toe, justifying here the use of linear stress extrapolation according to Eq. 4.2 for the 10 mm case.

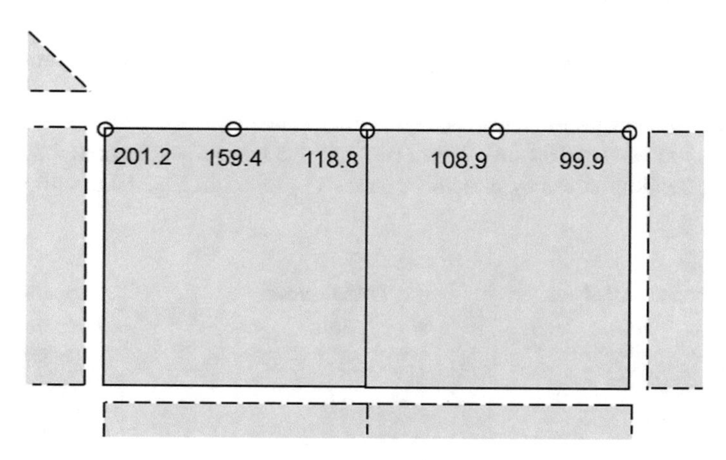

Fig. 10.3 Computed nodal stresses along the *upper* edge of the stiffener for 10 mm element length

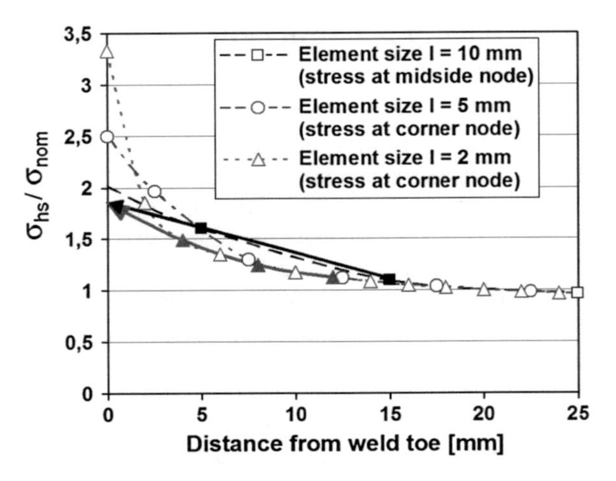

Fig. 10.4 Distribution and extrapolation of stresses close to the weld toe (Solid symbols show stresses used in extrapolation)

10.2.3 Determination of the Structural Stress According to Dong

The structural stress according to Dong [2] was determined for a linearized stress distribution over the assumed crack depth at the weld toe of 10 mm. This linearized stress distribution is computed from stresses at a distance δ = 10 mm from the weld toe in order to avoid influences from the stress singularity. The structural stress is obtained using equilibrium of forces and moments.

Figure 10.5 shows the stress components in the element next to the weld toe for the coarse model (l = 10 mm). These result in the internal stress components (axial stress σ_m, bending stress σ_b and shear stress τ) plotted in Fig. 10.6 with values for

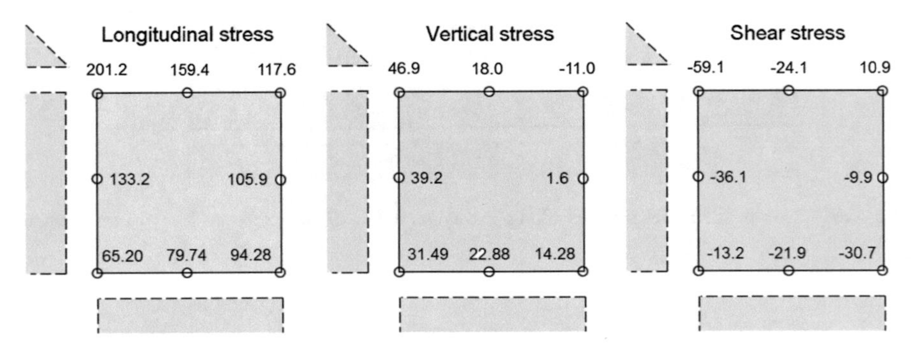

Fig. 10.5 Computed nodal stress components for the 10 mm long element next to the weld toe

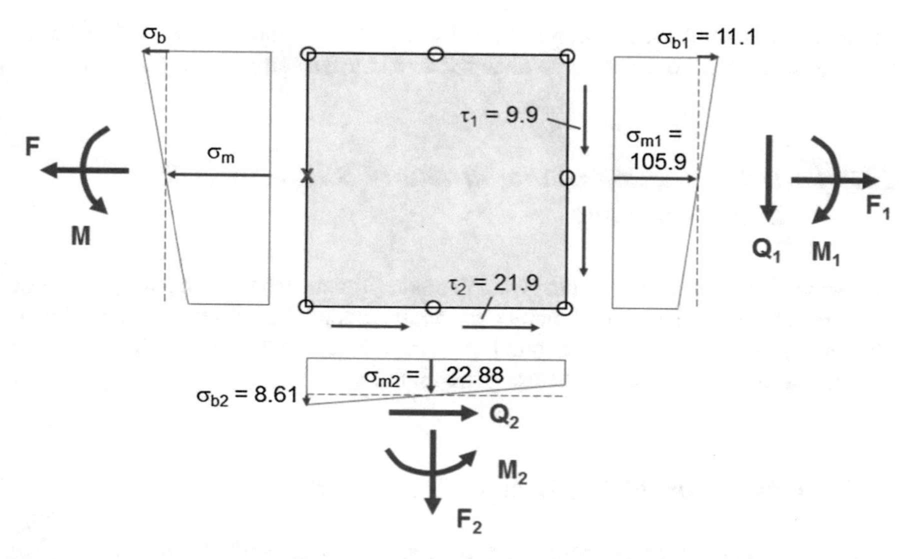

Fig. 10.6 Internal forces, moments and stresses acting at the 10 mm long element next to the weld toe

the right and lower edge. These can easily be converted into axial forces F, bending moments M and shear forces Q using the following equations:

$$F = \sigma_{\mathrm{m}} \cdot \ell \cdot t \tag{10.1}$$

$$M = \sigma_{\mathrm{b}} \cdot t \cdot \ell^2/6 \tag{10.2}$$

$$Q = \tau \cdot \ell \cdot t \tag{10.3}$$

where t is the element thickness.

The force F and moment M at the left side is obtained from equilibrium of horizontal forces

$$F = F_1 + Q_2 = 12,780 \text{ N} \tag{10.4}$$

and of the moment about the left mid-side node (indicated by “x” in Fig. 10.6):

$$M = M_1 - M_2 + Q_1 \cdot l - Q_2 \cdot l/2 + F_2 l/2 = 10,805 \text{ Nmm} \tag{10.5}$$

The corresponding axial and bending stresses at the left edge are

$$\begin{aligned} \sigma_{\mathrm{m}} &= 128.8 \text{ N/mm}^2 \\ \sigma_{\mathrm{b}} &= 64.8 \text{ N/mm}^2 \end{aligned} \tag{10.6}$$

so that the structural stress becomes σ_s = 192.6 N/mm^2. This stress is slightly larger than the structural hot-spot stress obtained from surface stress extrapolation.

10.2.4 Determination of the Structural Stress According to Xiao/Yamada

The structural stress according to Xiao/Yamada [3] was determined using the finest mesh model with 0.5 mm long elements from the nodal stress on the expected crack path at a depth of 1 mm. It was found to be σ_s = 195 N/mm^2, again slightly higher than that produced by surface stress extrapolation.

10.3 Estimation of the Design Fatigue Life

10.3.1 Fatigue Life Determined from Extrapolated Stress

FAT 90 is assumed according to Table 6.1 for the Type “b” hot spot. The corresponding design fatigue life according to Eq. (6.1) is as follows:

$$N = 2 \cdot 10^6 \cdot \mathrm{FAT}^3/\Delta\sigma_{\mathrm{hs}}^3 = 231,600$$

10.3.2 Fatigue Life Determined from Dong’s Approach

The structural stress parameter ΔS_s according to Eq. (6.3) is calculated assuming a 2D crack at the plate edge and the situation ‘load control’, resulting in $[I(r)]^{1/m}$ = 1.1 for the prevailing degree of bending $r = \Delta\sigma_b/\Delta\sigma_s$ = 0.33. The structural stress parameter becomes with m = 3.6:

$$\Delta S_s = \Delta\sigma_s \cdot t^{\frac{m-2}{2m}}/I(r)^{\frac{1}{m}} = 293^\circ \mathrm{N/mm}^2$$

where t is the depth of the assumed crack length (10 mm) in case of Type “b” hot-spots. The design fatigue life using Eqs. (6.3 and 6.4) becomes:

$$N = C/\Delta S_s^{m'} = 172,000$$

10.3.3 *Fatigue Life Determined from Xiao/Yamada's Approach*

FAT 100 is assumed as mentioned in Sect. 6.3.2. The design fatigue life according to Eq. (6.1) becomes:

$$N = 2 \cdot 10^6 \cdot \mathrm{FAT}^3/\Delta\sigma_s^3 = 269,700$$

10.3.4 *Comparison with Test Results*

From the fatigue tests [1], a design *S-N* curve was established yielding a fatigue life of N = 1,312,000 cycles for $\Delta\sigma_{\mathrm{nom}}$ = 100 N/mm^2 [4]. The fatigue lives determined with the structural stresses are much shorter, i.e. they are over-conservative.

Further investigations have shown that the critical weld toe was subjected to high compressive residual stresses induced by vertical shrinking during welding. These residual stresses are typical for Type "b" hot spots at welded attachments, whereas the Type "a" hot spots, i.e. the upper weld toe in Fig. 10.2b, usually exhibit tensile residual stresses [5].

The beneficial compressive residual stresses observed at the specimens [1] might, however, be relaxed during high loads or post-weld treatment like annealing. Additional fatigue tests of two post-weld heat treated specimens, relaxing the welding-induced residual stresses, showed fatigue lives reduced by a factor of 2.8 [6], agreeing much better with the predicted fatigue lives presented above.

References

1. Kim, W.S., Lotsberg, I.: Fatigue strength of load-carrying box fillet weldment in ship structure. In: Practical Design of Ships and Other Floating Structures (Ed. Y.-S. Wu, W.-C. Cui and G.-J. Zhou), Elsevier (2001)
2. Dong, P.: A Structural stress definition and numerical implementation for fatigue analysis of welded joints. Intl. J. Fatigue **23**, 865–876 (2001)
3. Xiao, Z.G., Yamada, K.: A method of determining geometric stress for fatigue strength evaluation of steel welded joints. Int. J. Fatigue **26**, 1277–1293 (2004)
4. Fricke, W., Kahl, A.: Comparison of different structural stress approaches for fatigue assessment of welded ship structures. Mar. Struct. **18**, 473–488 (2005)
5. Fricke, W.: Effects of residual stresses on the fatigue behaviour of welded steel structures. Mat.-wiss. u. Werkstofftech, 36 (2005), pp. 642–649 and Proc. of 1st Symp. on Structural Durability, Darmstadt (2005)
6. Kim, W.S., Lotsberg, I.: Fatigue test data for welded connections in ship shaped structures. J. Offshore Arctic Eng. **127**, 359–365 (2005)

Appendix

Symbols[1]

a	weld throat size
C	constant in equation of *S-N* curve with exponent m
C_1	constant in equation of *S-N* curve with exponent m_1
C_2	constant in equation of *S-N* curve with exponent m_2
D	fatigue damage $\Sigma(n_i/N_i)$ according to Miner's rule
D	mean diameter
D_{max}	maximum diameter
D_{min}	minimum diameter
d	deviation from true circle due to angular misalignment
E	elastic modulus
e	axial misalignment (eccentricity or centre-line mismatch)
f	function
F	force
FAT	fatigue class (characteristic fatigue strength at $N = 2 \times 10^6$)
h	weld leg length
I	fatigue life integral
K_m	stress magnification factor due to misalignment
K_s	structural hot-spot stress concentration factor
$K_{s,m}$	structural stress concentration factor to be applied to axial or membrane stress component
$K_{s,b}$	structural stress concentration factor to be applied to bending stress component
L	attachment length in direction of loading considered (see Figs. 2.3 and 6.2)
l	distance from axially misaligned joint to load or extremities of region of angular misalignment (shortest distance = l_1)
l	length of finite element
M	bending moment
m	exponent in equation of *S-N* curve

[1]Unless otherwise defined in text.

E. Niemi et al., *Structural Hot-Spot Stress Approach to Fatigue Analysis of Welded Components*, IIW Collection, DOI 10.1007/978-981-10-5568-3

m_1 exponent in equation of *S-N* curve for stress ranges above $\Delta\sigma_{th}$ ('knee point')
m_2 exponent in equation of *S-N* curve for stress ranges below $\Delta\sigma_{th}$ ('knee point')
N fatigue life in cycles
n exponent in thickness correction (Eq. 6.2)
n_i number of cycles in applied stress spectrum at stress range $\Delta\sigma_i$
p internal pressure
Q shear force
r degree of bending $[\sigma_b/(\sigma_m + \sigma_b)]$
R_{eH} yield strength
S_S equivalent stress parameter
t section thickness
t_{eff} effective thickness (Eq. 6.2)
t_{ref} reference thickness up to which *S-N* curve is applicable without the need to apply a thickness correction
y height of peaking due to angular misalignment
α angular change at misaligned joint
β geometric parameter used to determine K_m due to angular misalignment
γ partial safety factor (suffix '*f*' refers to loading, suffix '*m*' to fatigue strength)
ε_A strain measured by strain gauge A
ε_B strain measured by strain gauge B
ε_{hs} hot spot strain
ε_x strain acting in x-direction
ε_y strain acting in y-direction
$\Delta\varepsilon$ strain range $(\varepsilon_{max} - \varepsilon_{min})$
ν Poisson's ratio
σ stress (suffix '*m*' for membrane stress, suffix '*b*' for bending stress)
σ_{hs} structural hot-spot stress
σ_{lb} local notch stress $(\sigma_m + \sigma_b + \sigma_{nlp})$
σ_{nlp} non-linear stress peak at weld toe due to notch effect of weld
σ_{nom} modified nominal stress (suffix '*m*' refers to the membrane stress component, suffix '*b*' to the bending stress component)
σ_s structural stress
$\sigma_{1,2}$ maximum and minimum principal stresses
$\sigma_\perp$ stress component acting normal to weld toe
$\Delta\sigma$ stress range $(\sigma_{max} - \sigma_{min})$
$\Delta\sigma_{hs}$ structural hot-spot stress range
$\Delta\sigma_{R,L}$ constant amplitude fatigue limit
$\Delta\sigma_{th}$ stress range at 'knee point' where exponent of *S-N* curve changes from m_1 to m_2
τ shear stress
Φ angular position of weld in pressurised cylinder exhibiting ovality

Printed in the United States
By Bookmasters